Process Operations

PROCESS OPERATIONS

LESSONS LEARNED IN A NONTECHNICAL LANGUAGE

PB®

PENNWELL BOOKS

Disclaimer

The recommendations, advice, descriptions, and the methods in this book are presented solely for educational purposes. The author and publisher assume no liability whatsoever for any loss or damage that results from the use of any of the material in this book. Use of the material in this book is solely at the risk of the user.

PennWell Books, LLC
10050 E 52nd Street
Tulsa, Oklahoma 74146 USA

866.777.1814
sales@pennwellbooks.com
www.pennwellbooks.com

Publisher: Matthew Dresher

ISBN 9781955578066 (Hardcover)
ISBN 9781955578073 (eBook)

Library of Congress Cataloging-in-Publication Data Available upon Request.

Printed in the United States of America

1 2 3 4 5 26 25 24 23 22

Contents

Figures

List of Tables

Letter from the Author

I went to university from 1961 to 1964, and I learned nothing. It wasn't that I didn't try. Yet, I have been really successful as a process engineer in designing, operating, and—most importantly—troubleshooting refinery process problems for the past 58 years.

Most of the knowledge that I rely upon consists of things I've learned from troubleshooting home repairs, swimming pool maintenance, and air-conditioner problems. I am a real expert of any and all toilet hydraulic failures.

Also, I had a wonderful high school science education, which I apply all the time—like, PV = NRT and 34 feet of water = 1 Bar.

This book wouldn't be possible without the support of the *Oil & Gas Journal*. Throughout my career I have contributed many articles and the foundation of this book comes from the "Beyond back-to-basics: Process principles and concepts" article series.

I hope my book helps you in your job in the process industry.

Norm Lieberman
Chemical Engineer
norm@lieberman-eng.com

Dedication

To my family:

- Liz
- Arline & Bill
- Lisa Bug
- Joe & Julie
- Irene & Allen

They are my lubricants that reduce frictional drag, as I flow thru life.

Introduction

I began my refinery troubleshooting career as a child, by wondering and worrying about the ordinary things that other, more normal children take for granted—things such as the steam radiator heater in the bedroom I shared with my older sister, how a drop of water dances around on the surface of a smooth frying pan, and why the water level in our toilet bowl would rise and lower a bit on very windy days. As a young boy, I had very little purpose in observing these oddities of the world around us. I suppose, in retrospect, that my interest in water, fire, wind, and air, was an outgrowth of a lack of cell phones, email, television, and microwaves—things that did not exist in 1948, when I was 6 years old. But perhaps the lack of exposure to the "Age of Computers" permitted me to develop the skills that I will be explaining to the readers of this book in the following chapters.

The purpose of this book is to tell process technology stories relating to things that have happened to me. I'm mainly interested in processes such as:

- Combustion
- Evaporation
- Flow of fluids
- Corrosion
- Solubility
- Boiling
- Vaporization
- Condensation
- Heating
- Pressure
- Refrigeration
- Cooling

My real passion in life is *steam*. Many of my stories concern this theme:

- Steam reboilers
- Steam turbines
- Steam vacuum ejectors
- Reciprocating steam engines
- Stripping steam

Process operation and engineering are complex branches of human knowledge. The process operator or engineer designs, operates, and troubleshoots the following:

- Petrochemical plants
- Refineries
- Pharmaceutical plants
- Food processing facilities

- Distilleries
- Steam and power plants
- Waste water treatment
- Water purification plants
- Hazardous waste disposal units

Their work has a lot to do with these topics:

- Combustion
- Chemistry
- Plumbing
- Thermodynamics
- Corrosion
- Hydraulics: Fluid flow
- Vapor-liquid equilibrium
- Heat balance
- Process control

I've related my observations drawn from everyday life into stories that I have applied to process problems. The key to safety and profitability in process units is to understand how things function. My success as a process engineer is a consequence of the ability to solve process plant problems using simple methods and grade school calculations. This is not a result of my engineering degree. It's based on observations I have made in all kinds of circumstances:

- As a child in our kitchen in New York
- Caring for my swimming pool in Louisiana
- Repairing toilets in my house in Illinois
- Studying a steam radiator in my bedroom in Washington
- Trips on a river steamboat on the Mississippi River
- Considering how my fireplace draws in combustion air in Texas
- Watching swirling water in a river in California
- Flying a kite
- Operating a microwave
- Watching water boil in an old frying pan

Geared to young and seasoned professionals alike, *Process Operations: Lessons Learned in a Nontechnical Language* is designed to present a straightforward approach to mastering the principles and concepts all process engineers should be able to apply without the need of a computer. While simulations and models are useful for examining long-term operational issues, they cannot replace the dimension of human logic and reason required when tackling the array of complex—and sometimes life-threatening—situations that occur in process plants. Using experiences from my more than 58-year career in the process industry, *Process*

Operations: Lessons Learned in a Nontechnical Language provides approaches to understanding core process concepts in ways that will equip the engineer to walk out of an office into a plant and directly resolve process deficiencies via small operational changes or simple retrofits.

The book is setup to be an on hand reference for you. It is not necessary to read it in sequential order. You will find the format of each chapter to be the same. Each chapter begins with an introduction of the subject followed by a purpose. Once each topic is explained the chapter closes with a Lessons Learned section.

1

Applying Technology at Home

Nucleate Boiling

I had just bought a new mug at Walmart; it was smooth and shiny, with no scratches. I filled the mug with water and put it in the microwave for 5 minutes. Nothing happened—the water did not boil. Puzzled, I set the very hot mug on the counter. Nothing happened.

I dropped a tea bag into the "super-saturated" water. The water exploded out of the mug with tremendous violence. The rough surface of my tea bag had provided sites for nucleate boiling to be initiated.

My mom had two frying pans. One was a very old, scratched pan that she inherited from her grandmother. She also had a new, smooth, stainless-steel pan—a present from my dad. If I heated the old pan on the stove for 10 minutes and gently dropped a tablespoon of water on the hot surface, the water would flash to steam and be entirely evaporated in a second.

If I heated the new, smooth pan on the stove for the same 10 minutes and, again, carefully deposited a tablespoon of water on its hot surface, the water would dance around on the pan without boiling. It would slowly evaporate over a period of several minutes.

Strange to say, the hotter the new stainless pan got, the longer it required for the tablespoon of water to evaporate. But, for the old cast iron pan, the hotter it was, the quicker the water would evaporate.

High School Chemistry Lab

The main purpose of attending high school is to meet girls. A secondary objective is to learn to use "boiling stones" in the chemistry lab. If you forget, when you attempt to agitate a flask's contents with a stirring rod, the contents may explode in your face.

This often is not a problem. If the lab flask is old and scratched, nucleate boiling sites probably already exist. Boiling stones, which provide these points where vaporization can be initiated, are not needed.

When I worked as a unit engineer at a distillation tower, I obtained a sample of overhead product from the reflux drum. The product was a mixture of low-boiling, volatile, liquid hydrocarbons. I drew the sample into a new glass bottle.

It was a cold day (30°F), and the sample was also cold. Back in my warm office, I set the cold bottle on my desk. Nothing happened. I had thought the volatile liquid hydrocarbons would boil off gradually, as the sample warmed from 30°F to 70°F. But, after an hour, no boiling had occurred.

So, I dropped a thermometer into the bottle. And the entire contents of the bottle exploded in my face. As the liquid warmed, it became super-saturated. But, once a site (i.e., the thermometer) was provided to initiate boiling, the sensible heat of the super-saturated liquid was instantaneously converted to the latent heat of vaporization.

Now, when I draw a sample under similar circumstances, I always first drop a pebble into the bottle.

Heat Flux Limited Situation

The idea of heat flux in vaporization service of a pure component is the amount of heat (calories or BTUs) that can flow through a certain area (ft^2 or $meter^2$) in a certain amount of time (hours). The industry standard maximum number is 20,000 BTU/HR/ft^2. While this value has been accepted in the process industry for years, it's wrong, even though major process companies and contractors adhere to this design criteria. My story from Texas City in 1969 illustrates my point and, of greater import, explains how to overcome the "heat flux limit" cited above.

Butane Vaporizer

I had designed a heat exchanger to vaporize liquid butane using steam. There were two steam sources:

A. 30 psig steam (condenses at 250°F)

B. 100 psig steam (condenses at 330°F)

The butane (at 200 psig) would vaporize at 160°F. So, steam at either pressure would be able to vaporize the butane liquid into the plant's fuel gas system.

The new vaporizer had smooth, stainless-steel tubes. When I started the 30 psig steam flow through the tubes and the butane flow through the shell side of the vaporizer heat exchanger, the amount of the butane vaporized was only 70% of my design. Huh? This was bad!

So, I tried the hotter, higher-pressure 100 psig steam (330°F) instead of the 30 psig (250°F) steam. But the butane vaporization rate went down—just like my experience with the smooth frying pan! A flux-limited situation!

Overcoming Flux-Limiting Heat Transfer Rates

The next day, relying on my childhood observations with frying pans, I then did the following:

- Had the tube bundle extracted from the shell
- Had the exterior of the tubes sandblasted, to roughen their surface
- Returned the vaporizer to service

I was then able to achieve the design vaporization rate, using 30 psig steam.

Lessons Learned

It isn't that modern, digital, process technology is right or wrong; it's irrelevant to solving hydrocarbon process problems. It isn't that our university degrees, teaching differential equations, and process simulations are wrong; they're just irrelevant. It's the lessons we carry with us from everyday experiences that help us solve process problems.

Try to understand how your home air-conditioner works. Why is it that hot water heated in a microwave in an old, scratched teacup boils readily, but not so in the smooth, new cups that I just purchased at Walmart? Why does the convex side of a kite face into the wind, but the concave side of the kite faces away from the wind? Why do our toilets have a loop on the drain?

I still have no actual friends, outside of my work as a process engineer. My lack of an outgoing personality prevented me from advancing when I worked for American Oil for 16 years. But, at age 80, I own a house with seven bathrooms and I am still troubleshooting refinery and petrochemical plant problems, using the knowledge I acquired in a small apartment as a child in south Brooklyn in New York City in the 1940s.

2

Applying Technology at Work

Introduction

My knowledge that resulted in my first successful refinery project—the increase in the refrigeration system of the American Oil "Viscous Polypropylene Unit" in El Dorado, Arkansas—was based on my understanding of what happens when my home air-conditioner unit gets low on Freon. My debottlenecking of steam-heated reboilers by sandblasting the process side of the tube bundle was based upon an argument with my ex-girlfriend, Carol, who discarded an old, cast-iron frying pan in favor of a new, smooth, stainless-steel pan. I am forever solving multimillion-dollar process problems by applying the technology I have learned at home, rather than technology taught when I received my chemical engineering degree from Cooper Union College in 1964.

Purpose

The purpose of this chapter is to acquaint the reader with how I go about applying process technology to design or retrofit or troubleshoot process equipment that is used in both refineries and petrochemical plants. To some extent, I do use the principles taught at universities:

- Bernoulli's equation
- McCabe-Thiel diagram
- Thermodynamics
- Relative volatility
- Mollier Diagram
- Heat and material balances

But, to an equal amount, I rely upon the technology I have absorbed by troubleshooting process equipment for 58 years, as well as my even greater experience in fixing toilets, air-conditioners, natural gas burners, fireplace draft problems, and Pool pumps, among other things.

The following are examples of how I have applied technology to process plant problems.

Polymer Production—El Dorado, Arkansas

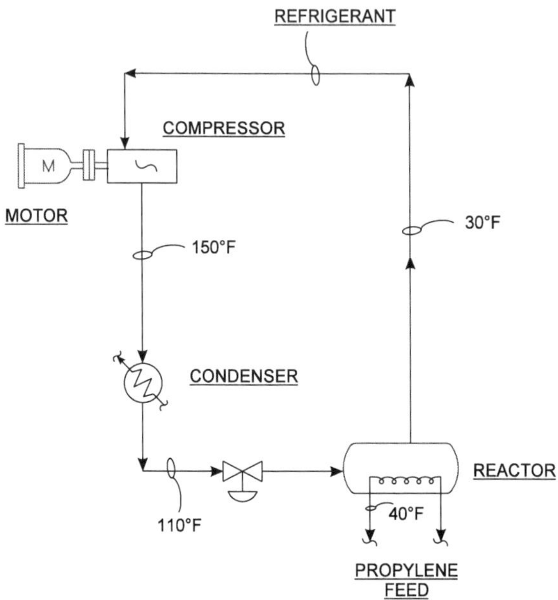

Fig. 2–1. Substituting a more volatile refrigerant increased refrigerant capacity.

- Limit—There was a reciprocating compressor in refrigeration service, driven by a fixed-speed motor, with spare amps. I needed to circulate more propane refrigerant. But I couldn't, because the volume of circulating propane refrigerant vapor was fixed, because the speed of the reciprocating compressor was fixed. That is, the alternating-current, three-phase motor was running at 3600 rpm.
- Background—This limit had existed for more than 10 years. The economics of the plant largely were dependent on the production of the polymer product.
- Solution—I gradually substituted the propylene feed for the current circulating refrigerant propane. The composition of the current circulating refrigerant was essentially 100% propane.

The composition of the feed, which was the depropanizer overhead stream from the fluid cracking unit, was as follows: 35% propane, 59% propylene, 6% ethane. The increased volatility of the feed, mainly due to the 6% ethane, raised the compressor suction pressure by 20%. This raised the density of refrigerant vapor at the compressor suction by 20%—which enabled the unit to increase the refrigerant circulation rate and hence increase the viscous polypropylene production by 20%.

- Follow-up—This change was only possible, because I was not limited by either (a) the refrigerant condenser maximum operating pressure, or (b) the amp load of the compressor motor driver. This 20% increase in production required three days to slowly alter the refrigerant composition. I remember with satisfaction that an older operator brought his son, who was studying engineering, to the refinery to explain what we had accomplished.

Vacuum Tower—New Orleans, Louisiana

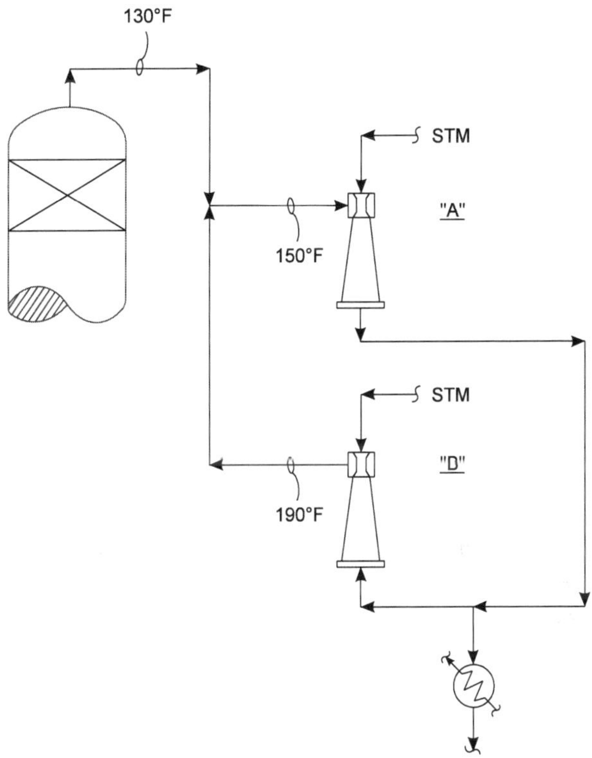

Fig. 2–2. Ejector "B" was flowing backwards

- Limit—There was a steam vacuum ejector that was pulling a vacuum from the top of a refinery fuel oil vacuum tower. The tower-top pressure was 35 mmHg, as compared to the design of 20 mmHg. The poor vacuum was reducing the amount of heavy gas oil that could be evaporated from vacuum tower tar. The tar had only 60% of the value of the gas oil.
- Background—The poor 35 mmHg vacuum had been normal for many years, and it had been accepted by operators and the management as okay, even though it was inferior to the design vacuum of 20 mmHg.
- Solution—The vacuum tower-top temperature was 130°F. There were two parallel, first-stage steam ejectors. I checked the process inlet

temperatures with my infrared surface thermometer and compared them to my observed tower-top temperature indication of 130°F. The inlet temperature to Jet "A" was 150°F. The INLET temperature to Jet "B" was 190°F (see Fig. 2–2). I reasoned that some of the motive steam to Jet B was flowing backwards to Jet A. I then blocked in the inlet valve to Jet B in this order: process valve first, steam valve second. Within five minutes, the vacuum had improved from 35 mmHg, to 22 mmHg.

- Follow-up—Recovery of the gas oil product from the tar product increased by 3,000 BSD within two hours. Also, steam consumption on the ejectors was halved by 4,000 lbs/hr. This change was worth about $50,000 a day to the Tenneco Oil Refinery.

Later, I found that Jet B's steam nozzle threads were corroded, as a result of the 316 stainless-steel nozzle being screwed into a carbon-steel ejector body, plus the fact that the steam was wet. The A nozzle threads had been wrapped with Teflon tape to suppress the effects of galvanic corrosion, due to wet motive steam, but not the B nozzle threads.

Note that this incident only occurred because both A and B steam ejectors vented to a common condenser. Also, a note of caution: When taking a steam ejector out of service, *always* block in the process valve first, to avoid unit upset of the process gas flowing backwards when the motive steam flow is stopped.

Pumparound Optimization—Norco, Louisiana

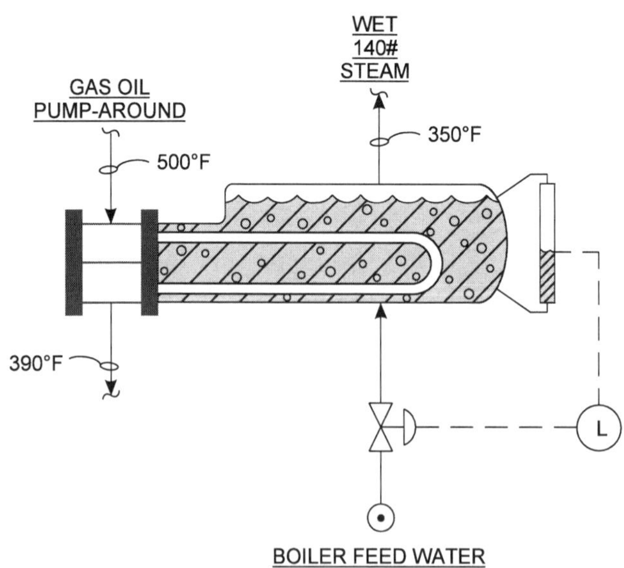

Fig. 2–3. Water level in kettle is higher than water level in glass.

- Limit—There was wet stripping steam to a gas oil stripper.
- Background—The crude tower gas oil product contained 35% diesel. Water in the steam had rendered stripping efficiency ineffective.
- Solution—The stripping steam was coming from a kettle-type waste heat boiler, extracting heat from the gas oil pumparound loop. In such a boiler, the indicated level in the gauge glass will always be lower than the water level in the kettle. The reason being that the water in the boiler is less dense (because it is boiling) than the cold water in the external gauge glass. Hence, while the set-point level in the glass was an inch higher than the top tube in the kettle, the actual level in the kettle was too high, and therefore, water was carrying over into the stripping steam.

To correct the problem, I lowered the level in the glass until the gas oil outlet temperature increased by a few degrees. This indicated I had uncovered the top row of tubes. I then raised the level in the glass by one inch. This dried out the steam.

- Follow-up—Diesel in gas oil was reduced from 35% to 20%, due to the drier steam. At $10/barrel, this was worth $3,000/day. Big bucks!

Air Blower—Norco, Louisiana

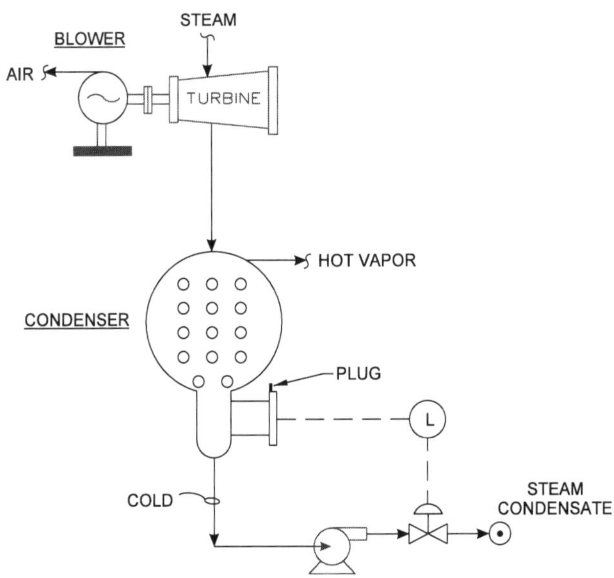

Fig. 2–4. Loose plug causes reduced turbine horsepower

- Limit—The fluid catalytic cracking unit (FCU) combustion air blower was running too slow, due to lack of turbine horsepower. This prevented the blower from developing sufficient discharge pressure to overcome back-pressure from the FCU catalyst regenerator. The cause of the reduced turbine horsepower had been identified by the unit operators as a bad vacuum in the surface condenser.

- Background—The steam turbine discharged into a water-cooled surface condenser with a normal operating pressure of 40 mmHg. The current surface condenser pressure was 110 mmHg. The increase in condenser pressure above the normal 40 mmHg had reduced the horsepower developed by the turbine by about 15% to 20%. Hence, the slower blower speed. I observed that the cause of the excessive turbine exhaust pressure was the excessive (130°F) surface condenser vapor outlet temperature. I also noted that the water boot temperature was a subcooled 80°F. Ordinarily, the vapor and the boot temperatures should be the same, unless the boot is flooded. However, I could plainly see the water level in the boot's gauge glass was only 50%.
- Solution—By accident, I felt that the plug on the top of the boot-level gauge glass was loose. As soon as I tightened it by hand, the following occurred:
    - The indicated level in the boot began to increase in the gauge glass.
    - The boot level control valve opened.
    - The condenser vapor outlet temperature dropped.
    - The condenser liquid outlet temperature went up to equalize with the falling vapor outlet temperature.
    - The vacuum improved to 40 mmHg.
    - The turbine speed increased.
    - And the air blower discharge pressure increased to normal.
- Follow-up—The owner of the refinery heard about this incident, and promoted me to tech service manager and, eventually, to plant manager.

Bad Fractionation—Convent, Louisiana

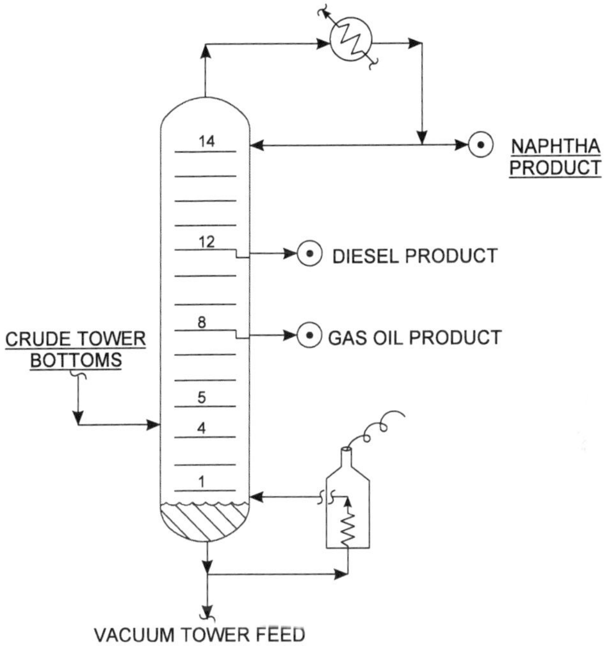

Fig. 2–5. Trays 8–12 had low fractionation efficiency due to low vapor rates.

- Limit—Bad fractionation between diesel and gas oil in a trayed fractionator resulted in half of the diesel in the tower's feed being downgraded to gas oil.
- Background—The tower was a 16'-0" I.D. fractionator, which had been greatly oversized. Increasing the fired reboiler duty was not an option, as the fractionator was limited by overhead condenser capacity. I measured the pressure drop across the trays between the flash zone and the top tray as essentially zero. My initial perception was that the trays were damaged and needed to be replaced. However, when the tower was opened for inspection two years later, I found that all the trays were intact.

  I then calculated, using operating conditions, the vapor pressure drop flowing through the valve caps. The calculated DP, based on the operating conditions, was about 0.2 inches of liquid. As the weir height was 3 inches, the valve trays were subject to severe tray dumping. This promoted bad vapor-liquid mixing on the tray decks, and hence poor tray fractionation efficiency.
- Solution—To correct the problem, I instructed the turnaround maintenance supervisor to cover 75% of the valve caps. This would reduce the

tray open area of the tray deck by 75% and increase the vapor velocity through the remaining holes (valves) from 4 ft/sec., to 16 ft/sec.

As pressure drop varies with velocity squared,

○ (DP) varies $(V)^2$

The calculated pressure drop of the vapor flowing through the tray deck increased from:

○ $(0.2) \cdot (16 \div 4)^2 = 3.2$ inches of liquid.

This was approximately equal to 3-inch weir height.

- Follow-up—After startup, fractionation efficiency between diesel and gas oil greatly improved. Virgin diesel oil production increased from 3,000 BSD to 5,500 BSD. Ken Starr, the operations manager, wrote a description in the refinery newspaper about the incident and referenced our company name. He also wrote me a nice note, which I have saved.

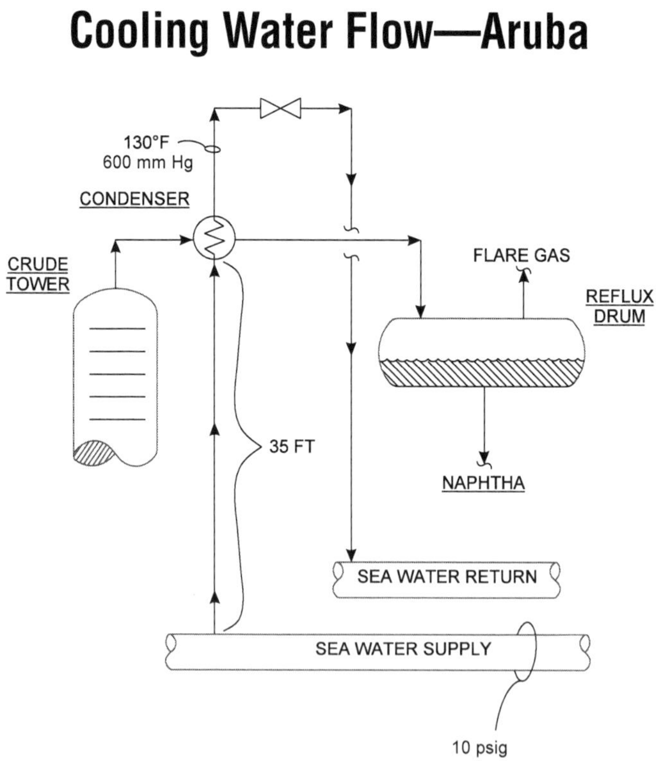

Fig. 2–6. Excessive elevation of a water cooler can cause air to flash out of water in the condenser tubes.

- Limit—A lack of cooling sea water flow to the elevated overhead condenser was causing excessive venting to the flare from the reflux drum.

- Background—Cleaning the sea water tube side of the exchanger had not helped. However, an operator mentioned that partly closing the sea water outlet valve reduced the flow of reflux drum off-gas to the flare.
- Solution—I monitored three points as I slowly, partially closed the sea water outlet valve from the condenser: water outlet temperature, reflux drum temperature, and condenser sea water outlet pressure.

  With the outlet valve wide open, I measured a water outlet pressure of 600 mmHg (i.e., 24 inches of Hg vacuum). At this subatmospheric pressure, and the 130°F sea water outlet, air flashed out of the sea water. The evolved air restricted the sea water flow through the condenser's tubes. Partly closing the water outlet gate valve had accomplished the following:

  - The water outlet pressure increased from 600 to 740 mmHg (atmospheric pressure in Aruba is 760 mmHg).
  - Air evolution from the sea water effluent was suppressed.
  - The sea water outlet temperature declined.
  - The reflux drum temperature also declined.
  - Off-gas flow to the flare diminished.
- Follow-up—Further closing of the sea water outlet valve increased both the water outlet, and reflux drum temperatures. The ultimate cause of this common problem is elevating the condenser too high. Then, at the top of the condenser, the water pressure falls several inches of mercury below atmospheric pressure and air flashes out of the water. This problem occurs with circulating cooling-tower water flow, as well as sea water.

Low-reflux Rate—El Dorado, Arkansas

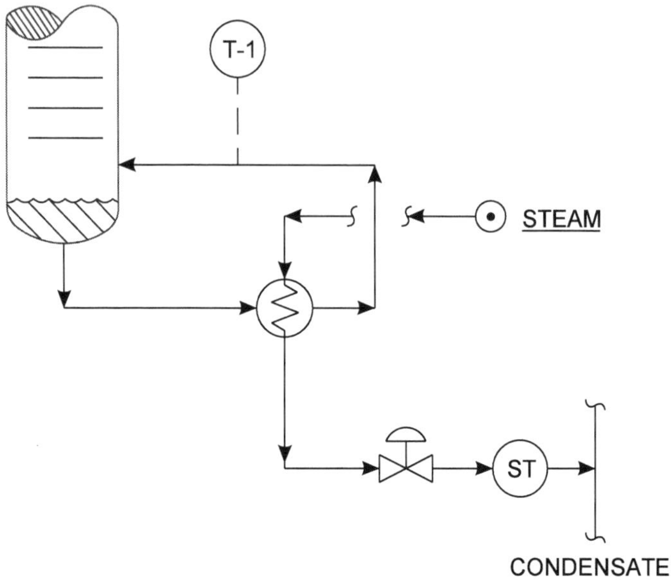

Fig. 2–7. Steam traps can either blow-thru or back-up. Either will reduce reboiler duty.

- Limit—Fractionation was poor because of low-reflux rates. When the reflux rate was raised, the reflux drum level declined, and the reflux pump lost suction.
- Background—The problem of the low-reflux rate was lack of reboiler duty. That is, the reflux comes from the reboiler vapor. Reboiler duty is low because either (a) team condensate is backing up, or (b) steam is blowing through the exchanger. That is, "loss on the condensate seal," due to a steam trap malfunction.
- Solution—To discriminate between these common problems, I slowly began to close on the steam condensate outlet flow. Simultaneously, using my infrared temperature gun, I monitored the reboiler process outlet temperature (T-1) (Fig. 2–7). If the problem was condensate backup, this would lower the process outlet temperature.

  If the problem was steam blowing through (i.e., loss of the condensate seal), this would increase the process outlet temperature at T-1.

  In this instance, the problem was improper maintenance of the condensate seal, due to a faulty steam trap, which was blowing through.
- Follow-up—Best to have a condensate drum and then pump the condensate to the collection system. Steam traps can either blow through, or fail closed, and cause condensate backup.

Spent Caustic Disposal—Texas City, Texas

- Limit—Disposal of spent caustic was a big expense. The caustic was used to extract H_2S (hydrogen sulfide) from a propane-propylene stream, which was one of the feeds to the alkylation unit:
  - $NaOH + H_2S \rightarrow NaHS + H_2O$
- Background—The spent caustic was stored in a large tank, the contents of which were:
  - 86% H_2O
  - 4% NaOH
  - 10% NaHS

  The alkylation unit shared the site with a spent sulfuric acid regeneration plant. One of the waste products from the regeneration of spent acid was 15% H_2SO_4 and 85% water (i.e., weak acid). This weak acid was stored in a large open-air pond. The contents of this pond were normally neutralized with purchased limestone, prior to discharge into Galveston Bay.

  One day, I had an idea to save the cost of imported limestone used to neutralize the waste sodium sulfide (NaHS). I phoned Bobby, my assistant, to start flowing the waste caustic into the weak acid pond.
- Follow-up—The following reaction then proceeded:
  - $NaHS + H_2SO_4 \rightarrow Na_2(SO_4) + H_2S$

  The resulting hydrogen sulfide cloud almost killed all the operators.

  Note: H_2S is fatal at a concentration of 1200 ppm or less. At this concentration, it has no odor. If you can smell H_2S, it's not at a deadly concentration. Three guys were killed on my Alky Unit in Texas City, two years before I assumed my superintendent position in 1974, by H_2S backing out of the flare system.

Lessons Learned

Most of the centrifugal pump and cooling water treatment problems I have faced were resolved based on my vast experience in maintaining my 48,000-gallon pool at my home in New Orleans. If you learn nothing else from my book, learn this: don't buy a house with a pool and seven bathrooms until you retire—which I haven't yet, at age 80!

3

Optimizing Fractionation Efficiency

Introduction

There is no such thing as a "fractionation tray." The purpose of the tray is never to fractionate. The tray deck is a vapor-liquid mixing device. The better the mixing, the better the tray's "fractionation efficiency." The vapor space above the tray is there to separate the droplets of entrained liquid from the rising vapors.

The downcomers serve the purpose of allowing the foam (a vapor-liquid mixture) to disengage into clear liquid and vapor, which bubbles back to the tray deck above.

Purpose

A big problem we have in tray efficiency is entrainment of droplets of liquid in the rising vapors. That's called jet flood (or incipient flood, or vapor flooding). Mainly, it is a consequence of excessive vapor velocity and excessive vapor density. Another problem that could happen is that the downcomers are undersized for the liquid flows, if the liquid is in the form of foaming.

The main purpose of this chapter is to emphasize the two principal problems in optimizing tray fractionation efficiency: *tray deck levelness* and *weir levelness*. Both of these determine vapor-liquid mixing efficiency of a distillation (or fractionation) tray.

Distillation

On my home planet, we have two types of molecules: blue and red. The blue molecules are light and lively. The red molecules are heavy and boring. The leader of our planet, Bossman, has developed a unique method to separate and segregate the light blue molecules from the heavy red molecules. The Bossman has christened this method "distillation."

It works like this: in a very tall, empty column, 50 perforated plates have been inserted. Heat is added at the bottom of the column (reboiler) and heat is extracted at the top of the column (condenser). Some of the condenser vapors run back down the column (reflux). And some of the liquid running down the column is vaporized (reboiler heat input).

Each plate has an overflow pipe (downcomer). The mixture of the blue and red molecules enters the column at plate 25 (the feed plate). Our Bossman has told us to adjust the top reflux rate and the bottom reboiler heat input to minimize the following:

- Blue molecules in the column bottom's red flow
- Red molecules in the column top's blue flow

I, our planet's engineer, have been given the above assignment.

Bubble Cap Trays

I suggested to our Bossman that we retrofit our column with bubble cap trays, as used on Planet Earth (a minor solar satellite in a nearby galaxy). But Bossman insisted that, on our planet, only perforated trays would be used. Why? He did not explain.

"Sir Bossman, perforated trays are subject to leakage. That is, the liquid drips through the trays and fails to efficiently contact the rising vapors," I explained.

"Engineer," Bossman said, "reduce the number of holes on the tray. Thus, the pressure drop of the vapor, as it accelerates through the holes, will stop the tray from leaking."

"But, Sir Bossman, how much velocity do I need to stop the tray from leaking?"

"You fool! Enough velocity so that the pressure drop of the vapor equals the weight of liquid on the tray."

"But, Fearless Leader! The trays are not completely level. Some parts of the tray deck are several inches below other parts."

Optimizing Reflux Rate

"Thanks to our Creator, I am here to guide you, Mr. Engineer," said Bossman. "What you must do is to extract more heat from the vapors leaving the top of the tower. This will generate more reflux. The reflux will flow down over the perforated plates and be re-vaporized in the reboiler. The new vapor flow will flow through the perforated trays, to stop them from leaking."

"But, Sir Bossman, I tried what you said. And it worked—up to a point. At first, the red molecules' contamination of the blue column top product diminished, as I increased heat extraction in the overhead. And I was happy! But then, more and more of the red molecules appeared at the top of the column. So, I tried to remove more heat from the column's condenser, to generate more reflux. But that only seemed to encourage the flow of red molecules to jump even higher and higher."

"You idiot. You went too far! You caused the reboiler vapor flow to increase so much that entrainment or flooding resulted!" said my Bossman.

Tray Weeping vs. Entrainment

"Yes, oh wise Bossman! You are correct! In my desire to increase vapor velocity to stop the trays from weeping, I caused the foaming vapors from the trays below to blow up to the trays above."

"Engineer, you should have calculated how much vapor you needed to stop the perforated trays from leaking or weeping."

"Ah, Sir, but when the trays were installed, I forgot to record how much they were 'out-of-level.' Without this information, I cannot see how such a calculation can be achieved."

Tray Pressure Drop vs. Flooding

"Praise the Creator who placed me in charge of this planet. Let my wisdom be a guide to you. What you should do is to measure the pressure drop across our column. When the pressure drop, expressed in inches of liquid, exceeds 18% to 25% of the column height, then you will contaminate the blue, light product with the red, heavier product, due to entrainment or flooding," explained Bossman.

"When the pressure drop, expressed in inches of liquid, is 10%-15% of the column height, then you will contaminate the blue, light product with the red, heavier product, due to tray deck leakage or dumping."

"But, Bossman, suppose it's only the top few trays that are suffering from entrainment. Then, the column pressure drop may be quite low, but the red molecules may be entrained overhead anyway?"

Tower Temperature Profile

"Engineer! Then you can just be guided by the temperature difference between the top of the column and the bottom. As you increase the reflux rate and the reboiler duty, the temperature difference should increase. The blue molecules are always cooler than the red ones. If you increase the reflux rate—and the reboiler heat—the column top temperature must go down, and the column bottom temperature must go up. Unless the trays are suffering from entrainment."

"But, Sir. Suppose it's only the top few trays in the column that are suffering from bad entrainment. Then, the column top temperature could fall suddenly. And the overall temperature difference in the column will get bigger. But, at the same time, the blue overhead product will get more red?

"Bossman! I do not think, based on my experiences, that either the column differential pressure or the differential temperature is a good guide for either keeping the red molecules out of the blue product going overhead or keeping the blue molecules out of the red bottom product."

Color Discrimination

"Engineer, go forth and sample the overhead blue product. If it contains more red molecules when you increase the column reboiler heat input and the condenser heat extraction, then the bad separation between blue and red is a consequence of flooding, and less reflux is needed.

"Sample the bottoms product. If it contains fewer blue molecules as you reduce the reboiler heat input, then the separation efficiency between blue and red molecules is bad due to entrainment or flooding, and less top reflux is needed.

"Be guided by sample results, and disregard both the temperature difference and the pressure difference across our column."

Closed-Loop Color Control on Earth

I have used closed-loop color control for drawing off atmospheric gas oil (AGO) product from a crude unit. The panel operator would ask the outside operator to show him a sample of AGO every few hours. If the AGO was brown, he would cut back flow of the AGO product just a bit. If the AGO was golden-yellow, he would increase AGO flow.

One day, an online colorimeter was added. Then, every once in a while, the panel operator would glance at the colorimeter and make a slight adjustment to the AGO product flow.

My contribution was to close the loop. If we are separating blue and red molecules, and we had an online colorimeter at the top and bottom, we could use them to continuously optimize both the reflux rate and the reboiler duty. And we could ignore Delta P, Delta T, and computer control optimization.

Direct analyzer control is not new. It is commonly used to control not only color, but viscosity, boiling points, flash point, gravity, Reid Vapor Pressure (R.V.P), and a wide range of compositions. I have used it myself to control the ethane content of liquefied petroleum gas (LPG) at the Chevron refinery in Salt Lake City, Utah.

In these circumstances, the usual method is to control top reflux and/or steam flow to a reboiler, with an online gas chromatograph.

Gas Chromatography (G.C.) Closed-Loop Control

Why is G.C. closed-loop control not in widespread use in the petrochemical and refinery industries? There are several problems:

- Cost
- Maintenance
- Reliability
- Complexity of sample station
- Excessive response time (12 to 24 minutes)
- Installation complexity

If only we could look at a sample of the overhead and see if the blue color was tinged with red. Or look at a sample of the bottoms product and see if the red color was tinged with blue—and, as if by magic, determine the exact color contamination concentrations of each, at the very moment the changes were occurring.

Wouldn't that be wonderful if such an instrument actually existed?

Spectroscopy as a Sensor

A practical method for the use of closed-loop analyzer control is the use of spectroscopy technology. This instrument is relatively simple to install, requires little maintenance, and is less costly than a gas chromatograph (G.C.).

More importantly, the lag time between analysis and transmission of the composition of a product flow is reduced to a few seconds, rather than the many minutes required for G.C. I have seen such spectrometers in aromatic fractionation service, provided by Tec5USA. I rather believe this analyzer control will be the future of refinery distillation in light hydrocarbon services.

Fractionation Tray

These are the three functions of a tray in a distillation tower:

1. Completely mix rising vapor flow with descending liquid flow
2. Remove all droplets of liquid that are entrained in the rising vapor flow leaving the tray
3. Remove all bubbles of vapor from the foamy liquid leaving the tray

If these objectives are completely reached, then the tray will have 100% "fractionation tray efficiency."

In practice, none of these three objectives can be entirely achieved. Therefore, there can never be 100% fractionation tray efficiency.

Lessons Learned

If you really wish to learn important process lessons from my book, write this down: "Don't Trust Strangers." That is, crawl across the trays yourself during the turnaround:

- Are the outlet weirs at the correct height (typically 2–2½ inches)?
- Are the weirs level?
- Are the downcomer bracing brackets installed?
- Is the bottom edge of the downcomer level and just a little (½ inch) below the top edge of the outlet weir?
- Are the tray decks really level?

4

Reflux in Distillation

Introduction

The reflux washes back the heavier components in the ascending vapors. The original "stills" did not have reflux. To fractionate, both distillation trays and reflux are both required. Typically, the ratio of reflux to the overhead distillate product is 2–4 to 1. Without reflux, tray efficiency must be zero. Reflux is good—up to a point. The point is called "jet flooding." If we exceed the point of flooding (which I call the "incipient flood point"), then more flux hurts rather than aids fractionation efficiency.

Purpose

The main concept to understand distillation is that the reflux comes from the reboiler (and, to a lesser extent, from the vaporized portion of the feed, if there is a feed preheater). The archaic stills had no reflux. They simply vaporized a portion of the volatile alcohol in the "mash." The idea of reflux was developed in Cornwall in the 1830s for what was called the "patent still."

"Henry, why are you increasing reflux on the debutanizer?"

"I'm increasing the reflux because there's excess pentane in the butane overhead."

"Okay. But how about the reboiler duty? The steam flow to the reboiler is on manual. You'll have to increase the steam."

"No, Norm. I only have 2% butane in the bottoms product. It's right on spec. I don't need more reboiler steam. I'm okay with that."

"Henry, you can't just crank up the reflux rate without also increasing the heat to the reboiler. I'll explain it as I explained it to Jay, the plant manager of a refinery in Indiana, to help him with his crude unit and preflash tower."

"Norman," Jay said, "the project you suggested has been a big disappointment. I've diverted the crude flash drum vapors to the new condenser we installed in the last turnaround (Fig. 4–1). Diverting the 350°F flash drum vapors from the crude tower flash drum has increased flash zone temperature in the crude unit by 30°F.

Also, we've unloaded the crude tower trays and have been able to increase the crude rate by 7,000 BSD."

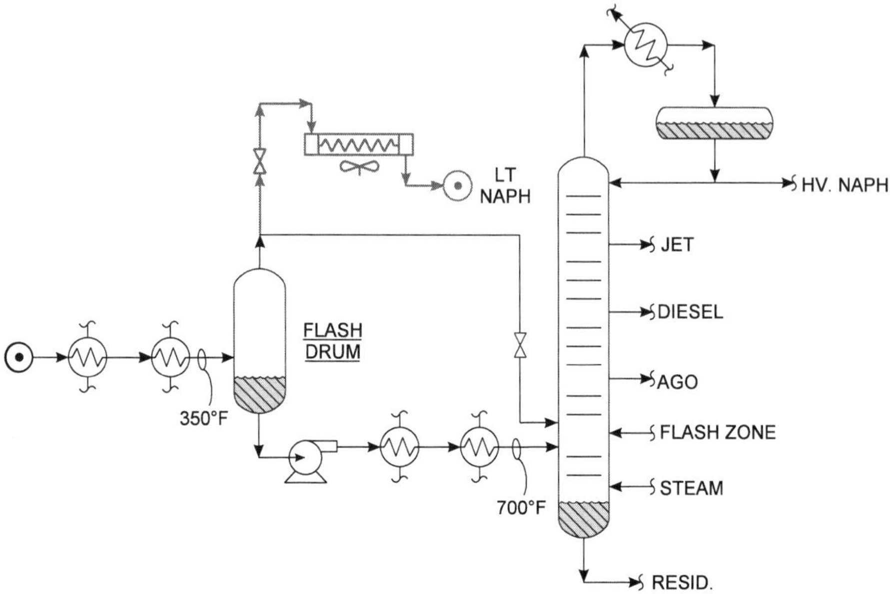

Fig. 4–1. Light naphtha has an excessive end point.

"Jay, that all sounds good. What's the trouble?"

"My trouble is that the light naphtha from the new flash drum vapor air cooler has a long tail. It has a 420°F end-point. That heavy tail is coking up my naphtha reformer catalyst. I need to get the ASTM D-86 Distillation end-point below 380°F max." (The ASTM D-86 test is the standard refinery test for gasoline, kerosene, and diesel.)

"Jay, I was visiting Peru a few years ago. On the road to Machu Picchu, a famous tourist destination, my driver suggested we stop for a tour at a distillery that made "Pisco Sour," the national drink of Peru."

I had drunk Pisco Sour at our hotel in Lima and thought it would be interesting to see the distillery, particularly as this was in the remote countryside.

I was introduced to the lead operator, Juan. Hundreds of clay jars were piled in one corner of the site. Stacks of firewood were in another area.

The still was a huge copper pot. It was set on a brick structure above a brick chamber into which firewood was being fed. A long coil of copper tubing condensed the vapors from the still, which drained the condensed Pisco vapors, into a vertical copper pot. This second pot was elevated 16 feet above the top of the still (Fig. 4–2).

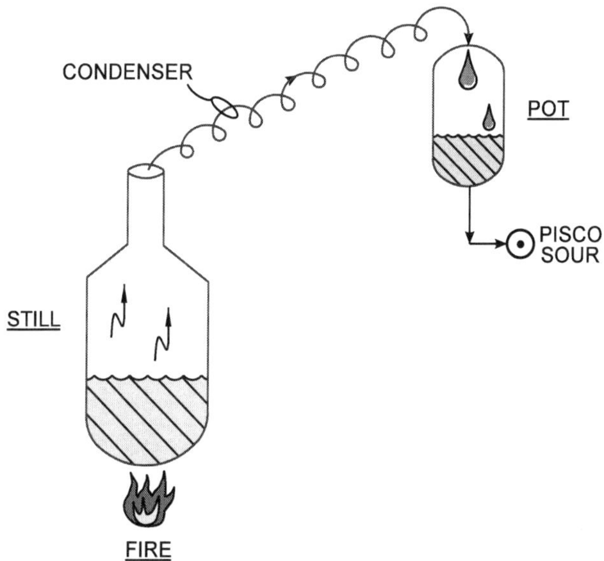

Fig. 4–2. Ancient still. Yield reduction for increased product purity.

Pisco is 25%–30% alcohol. Whiskey is 50% alcohol. At this distillery, they measured the alcohol content of the Pisco with a float. If the Pisco was too weak, the line on the float would be above the liquid level in their test glass, indicating too high a water content. Meaning the Pisco density was too high.

"So, Juan, what do you do to bring the Pisco up to 30% alcohol, if it's too weak?" I asked.

"Next batch, we don't cook it so hard. Use less wood. So, juice, it cooks slower and colder. Mucho caliente makes Pisco too weak. Or maybe, Señor, I put weak Pisco back into the still, heat it up a second time. The second trip through our still will make Pisco muy strong," Juan explained.

For me, this is what process engineering is all about. It's when I meet the plant operator, and we resolve our understanding of a problem using applied technology.

Vapor Pressure Limitations

"Juan," I asked, "Why do you sometimes cook the juice too hard if it makes weak Pisco? If it vaporizes too much water with the alcohol?"

"Because, Señor Norm, we make more Pisco product."

It's the original problem of process engineering. The balance between product recovery and product purity. To keep his product on spec, at a minimum of 30% alcohol, Juan had to reduce his product recovery, by burning less firewood underneath the still.

The problem is that alcohol has a large vapor pressure, and water has a smaller vapor pressure. When the still is fired harder, the vapor pressure of the water increases faster than the increase in vaporization rate of the alcohol from the fermented juice. So while the amount of alcohol vaporized increases, the amount of water vaporized increases even faster, because the vapor pressure of the water is increasing at the higher temperature in the still.

Reducing the still temperature reduces water vaporization, but also reduces the production. That's the price paid for making the Pisco stronger, to meet his alcohol spec. It's a compromise between product yield and product purity.

I explained the concept of vapor pressure to Juan, but he began to show disinterest. So I said, "Maybe we can fix the problem by using process technology."

Two-Stage Distillation Process

"Señor Juan, we are going to create a two-stage distillation process using a partial condenser. It will run continuously and process the Pisco all the time on spec—and cost almost nothing to install. We can do it today."

When I said that we could do it today and that it would cost nothing, Juan smiled and said, "Bueno!"

- Step 1—I'll cut the long coil of copper tubing in half.
- Step 2—The first half I'll use to connect the still to the vertical pot (Fig. 4–3).
- Step 3—The second half I'll leave connected to the top of the vertical pot and have it drain, by gravity, into one of those clay pots.
- Step 4—With the liquid that condenses (called "flux"), which accumulates in the vertical pot, I'll use a rubber hose with a 2-foot-long loop, to continuously drain back into the still (Fig. 4–3).

"Señor Juan, the liquid that drains back into the still will be re-vaporized in the still and is called flux or 'reflux.' The purpose of the loop is to keep vapors from the still from backing up into the clay pot's liquid drain line, which would stop the liquid draining through the hose. That is, the vapor would push the liquid back into the pot.

"But, Norm, suppose the Pisco is too weak. What should I do then?"

"Spray some water on the copper tubing between the still and the vertical pot. That coil being like a first-stage condenser, spraying water on the coil to remove more heat will make more flux, but less product.

"You see, Juan, I'm guessing about reducing the size of the existing first-stage condenser (Fig. 4–3). Maybe it should be a little bit longer to generate the right amount of reflux. But using the water spray will allow you to compensate for lack of cooling by air."

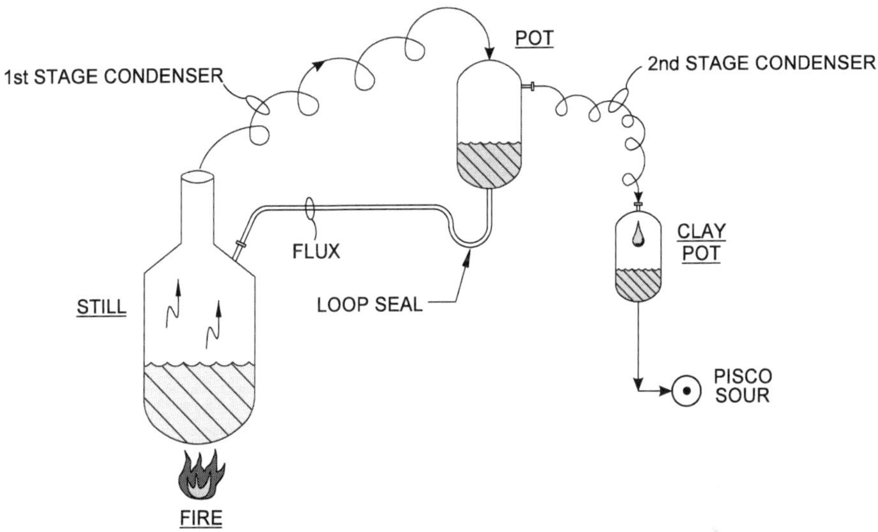

Fig. 4–3. Two-stage distillation apparatus increases purity without reduction in product yield.

"But, Señor Norm, suppose the Pisco becomes too strong! What should I do then?"

"Juan, just add more firewood to the still. That will make more Pisco product too, by distilling over some extra water."

Origin of a Distillation Column

The technology of the partial condenser to produce a strong alcoholic beverage originated in Persia around 1000 A.D. The multi-stage, trayed distillation tower that we use today originated in Scotland in the eighteenth century. It's called the "patent still." It had a reboiler, condenser, six bubble cap trays and a feed preheater.

The only things that have changed in distillation since the eighteenth century are that what we once called "flux" we now call "reflux," the still now has a separate reboiler, and we use valve trays instead of bubble-cap trays.

Removing the Naphtha's Tail

"Henry, we can do the same thing on your unit. Shut off several fans on the air cooler. Then, we'll take the liquid that was partially condensed in the air cooler outlet header box and drain it back into the flash drum, like reflux. The remaining naphtha vapor, we can route to the overhead of the crude tower (Fig. 4–4) and comingle it with crude tower overhead naphtha. The air cooler will be like the first-stage condenser in Peru, which is also working like a theoretical, fractionation stage."

"But, Norm, I'm going to wind up with less naphtha from the pre-flash tower, by using the air cooler as a partial condenser."

"It's a balance between product yield and product purity."

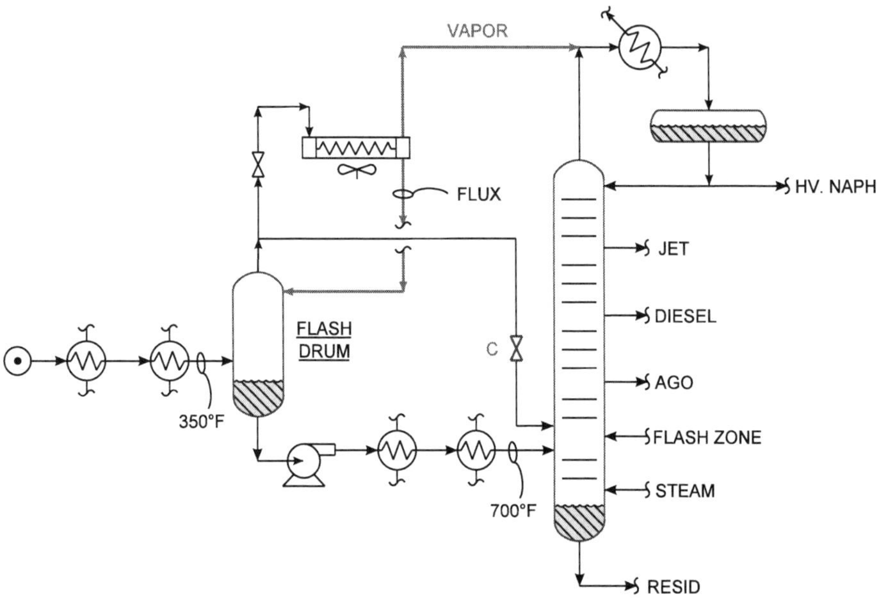

Fig. 4–4. The air cooler is acting as a theoretical separation stage.

"You certainly are quite the world traveler. Imagine yourself working in Peru," Henry said.

"Yes, Henry. Just like Jay's story, the travelling has its good and bad points. But the point of this story is that *the reflux comes from the reboiler*. If you increase the tower-top reflux flow, to reduce the pentane in the butane, the reflux drum is going to go empty, unless you refill it with vapors generated from the steam reboiler. The reflux pump is going to lose suction and cavitate, and then you'll tear up the pump seal."

"I'll remember your story and add steam to the reboiler next time I add reflux to the debutanizer."

Lessons Learned

- More reflux helps fractionation efficiency up to the “incipient flood point,” also called “jet flood.”
- The reflux comes from the reboiler and, to lesser extent, from the feed preheater.
- Increasing a distillation tower’s reflux rate, without more heat input, will cause the reflux drum to empty and cause the reflux pump to cavitate.
- There is no purpose in increasing the reflux rate if there is no ability to increase tower heat input and/or the tower’s trays are suffering from entrainment of “jet” flood.

5

Understanding How Distillation Trays Work

Introduction

The liquid flows through the downcomer to the tray below. It flows across the tray deck, which is perforated. Sometimes these perforations are called sieve holes. Sometimes they are called valve caps. And sometimes, bubble caps. Bubble caps are best, but rarely used in the industry. Valve caps—which are EVIL—are now the industry standard, but are gradually being replaced by "grid trays," which I like, but not as much as the archaic bubble cap trays.

The distillation trays' only function is to mix the vapor and liquid together. If the mixing—and subsequent separation of vapors and liquids—was perfect, the tray efficiency would be 100%.

Purpose

My main purpose in this chapter is to encourage the process engineer to design new, or modify existing, trays, so that the pressure drop of the vapor phase flowing through the tray deck is at least 50% of the weight of the liquid on the tray due to the height of the weir, plus the crest height of the liquid overflowing the weir. Of course, if the weirs and the tray deck panels are not level, then the calculated head of liquid on the trays, and the pressure drop of the vapor flowing through the tray decks, is meaningless.

A distillation tray is used to mix vapors and liquids together. The better the mixing, the more efficiently the tray works. Tray efficiency also depends on two other factors:

- How well the droplets of entrained liquid are separated from the vapor flow rising from the tray deck
- How well the bubbles of vapor are separated from the liquid flowing down from the tray deck through the downcomer

Entrainment and Tray Efficiency

As a young man, I would dream of meeting my heroes: Nikola Tesla, Daniel Bernoulli, Isaac Newton, and Sir George Stokes. Once, I dreamt that I had passed out of life, and was welcomed by, my famous colleagues.

"Welcome to the Eternal Life," Stokes announced to me cordially. "Can we answer any questions?"

"Mr. Stokes, I'm rather interested in entrainment. On Earth, this is a big problem affecting fractionation efficiency. Can you tell me, what factors increase entrainment?"

Stokes explained several factors that impact entrainment. Included were these factors:

- Droplet size. The smaller the drop, the greater the rate of entrainment.
- Viscosity of the continuous phase. High viscosity promotes entrainment.
- Density difference between the vapor and liquid phases. Reduced density difference promotes entrainment.
- Velocity. Higher velocities promote entrainment as well.

"I've actually reduced all these factors into a simple equation," Stokes explained, with a smile. "I call it Stokes' Law."

Mathematically, we say entrainment is proportional to:

$(V) \cdot (D_V \div D_L)$

where:

V = Velocity of vapor phase

D_V = Density of vapor phase

D_L = Density of liquid phase.

"But, Mr. Stokes, how about viscosity and the droplet size?"

"Well, Norman, the droplet size is usually unknown. Viscosity of a vapor phase is usually small, and I normally neglect it. By small, I mean less than 0.3 centipoise."

"That's all quite interesting. But if I could take a single step to suppress entrainment, what would you recommend?"

"The best option would be to avoid high-localized velocities of the vapor phase. Norman, I would also—"

"Wait a bit, Mr. Stokes!" Nikola Tesla interrupted. "Mr. Lieberman has arrived in Eternity prematurely. He's being deported back to his earthly home in New Orleans."

Entrainment

Originally, the process of distillation did not utilize trays. It relied on pots (Fig. 5–1). The pots, being quite large, completely separated the liquid phase from the vapor phase, before the liquid and vapor left each pot. That is, there was no entrainment.

Each pot was an individual separation stage. If the separation of vapor and liquid phases were perfect, the pot had a separation, or stage efficiency, of 100%. The more pots that were arranged in series, the greater the degree of separation that could be achieved between the two products.

In our case, we are trying to separate propane and butane as efficiently as possible. To avoid the expense and the mechanical complexity of dozens of these pots, the distillation column was invented about 180 years ago. Each tray is intended to represent a pot—or at least a portion of a pot. If the tray really is working as well as a pot, it is said to have 100% tray fractionation efficiency. If the tray is only working half as well as the pot, it is said to have a 50% tray efficiency. The usual cause of low tray efficiency is entrainment due to excessive vapor velocity.

As the distance between the liquid level on the tray below and the tray deck above gets smaller, it takes less vapor velocity to blow the liquid from the lower tray up to the tray above. For example, let's say that I am separating propane from butane. My propane product (e.g., LPG in 5-gallon containers sold in Walmart) has a maximum butane spec of 2%. If the vapor rate flowing up through a tower increases, the droplets of liquid from a lower tray will be entrained and blown up to the tray above. As the liquid in the lower tray is rich in butane, the entrained droplets will increase the butane content of the upper tray. Eventually, as the entrained droplets reach the top tray, the overhead LPG propane product will be contaminated with butane. Thus, fractionation efficiency is reduced.

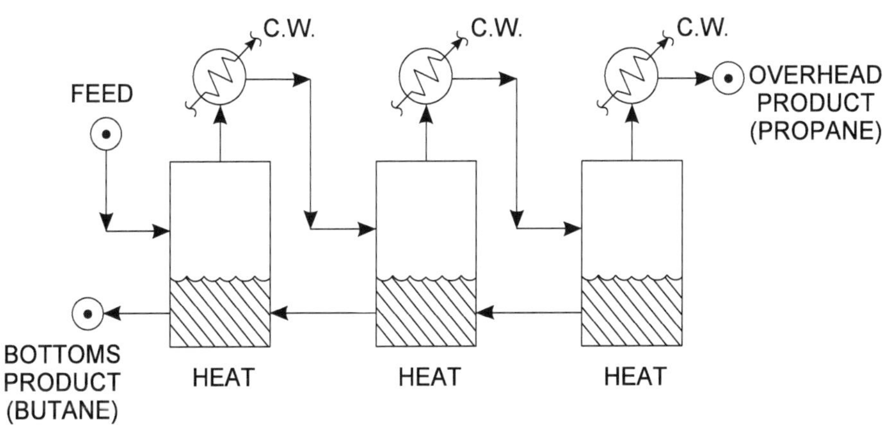

Fig. 5–1. Primitive distillation units three stages

Foam

The fluid on a distillation tray deck is not actually liquid, but foam or froth. The vapor bubbling up through the holes on the tray deck mixes with the liquid from the tray above, to form foam. The foam overflows the low dam (weir) into the downcomer (Fig. 5–2). If the downcomer is too small, however, the foam fills up the downcomer and backs up onto the tray above. The foam level on the tray then increases. As the froth, or foam level, increases on the tray, so does the height of the entrained liquid, or spray height. When the spray height from the tray below reaches the tray above, tray fractionation efficiency becomes worse due to entrainment.

Reflux

More reflux improves tray fractionation efficiency. The trays work harder and better due to the increased vapor and liquid traffic on the tray decks.

The problem is that the reflux comes from the vapor that is generated by the reboiler. If there is too much vapor flow through the tray decks, then the vapor velocity may become excessive. And according to Mr. Stokes, this will cause excessive spray height, or entrainment. It's rather like exercise. Too much can actually be bad for one's health. Too much reflux and too much reboiler heat are going to make fractionation efficiency worse, rather than better, due to entrainment.

You can't just increase the reflux rate without more reboiler duty, because the reflux is just condensed vapor generated from the reboiler, which rises up through the trays, through the overhead condenser, and into the tower's reflux drum.

Downcomer Backup

Downcomers often back up onto the tray above and cause a high-foam level on the tray—and a consequent entrainment of droplets of liquid to the tray above. Usually, it's not the downcomer itself that has too small an area. It's typically because the downcomer is too short. The height of foam in the downcomer is mainly a consequence of the pressure drop of the vapor flowing through the tray above the downcomer (Fig. 5–2). As the vapor flows through this tray, the vapor undergoes a pressure drop due to (a) the restriction of the caps, or orifices, on the tray deck; and (b) the weight of the liquid on the tray deck.

The pressure of the vapor below the tray pushes up the foam height in the downcomer. Once the downcomer becomes full of foam, the foam backs up onto the tray above that downcomer. The extra weight of foam on the tray deck then increases the pressure drop of the vapor through that tray deck. This, in turn, increases the force pushing up the liquid in the downcomer. The problem feeds upon itself, until the trays above where the problem began start to operate in fully developed flood.

A tray downcomer can also fill with foam because it can't drain out of the downcomer fast enough. This common problem is most often caused by a too-small downcomer clearance (Fig. 5–2). A reasonable downcomer clearance is usually 2 to 3 inches. Any clearance less than 1½ inches is likely to result in downcomer flooding and liquid backup on the tray above.

If the downcomer clearance is greater than the tray's outlet weir height, the downcomer will also back up onto the tray above. This is called unsealing of the downcomer. Vapor will blow into the bottom of the downcomer and displace the liquid flow, resulting in flood.

Keep in mind that the downcomer is full of frothy foam and not the denser liquid phase. This foam is easily displaced by the vapor flowing up through an unsealed tray downcomer.

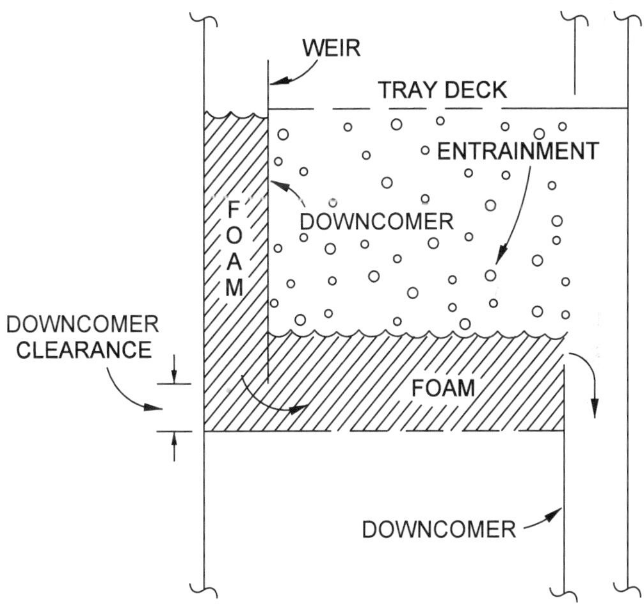

Fig. 5–2. Distillation tray entrainment reduces tray fractionation efficiency.

Tray Efficiency

Tray fractionation efficiency is reduced by entrainment, also called jet flood. Most tray efficiency problems are caused by tray decks that are not level. The liquid on the tray deck may leak through the perforations on the trays, in those areas which are lowest. The vapor flow will be concentrated at the highest tray deck areas. This all results in vapor-liquid channeling, which is bad for tray vapor-liquid contacting efficiency.

A tray that is out-of-level will cause a loss in tray fractionation efficiency and is related to the pressure drop of the vapor flowing through the tray deck holes or valve caps. If the low points of a tray (i.e., portions of the tray deck below the tray support ring) are less than the pressure drop of the vapor flowing through the tray deck perforations, then the liquid will leak through the tray deck and fail to come in good contact with the vapor flow.

Benefit of Valve Cap Trays

Valve caps are supposed to act as check valves to retard liquid leakage through tray decks. This appears not to be the case. The valves do not appreciably reduce tray deck leakage. What they certainly do in refinery dirty services is to stick to the tray decks. This restricts vapor flow and pushes up the foam level in the downcomer, which causes the tray to flood.

My practice is to use a grid-type tray. This is like a valve tray, but the cap is in a fixed position. It's more or less the same as a sieve tray, but with about 10% to 15% more capacity.

To calculate the pressure drop of the vapor through the tray perforations, use the following equation:

- $DP = K \cdot (D_V \div D_L) \cdot V^2$

where:

- DP = Inches of liquid
- D_V = Density of vapor
- D_L = Density of liquid
- V = Hole velocity, feet/second
- K (Sieve tray) = 0.3
- K (Valve tray) = 0.9
- K (Grid tray) = 0.4 to 0.7

Valve caps do not appreciably retard the tendency of liquid to leak through the tray deck panels. Bubble cap trays would retard the liquid leakage through the tray decks, but are unfortunately seldom used in a modern distillation tower. ▪

Safety Note: Inspecting Refinery Distillation Towers

Most refinery distillation towers will be fouled with iron sulfide. The black, fine material, which you find hard to wash off your hands, is not normally coke, but iron sulfide.

Iron sulfide will not autoignite until it dries. When dry, it will begin to burn at ambient temperatures and tend to smolder rather than flare up. The heat liberated is small. The sulfides burn with a pale blue flame, not visible in sunlight.

The danger is not heat, but SO_2 liberation in a confined space. Sulfur dioxide is deadly at 1,000 ppm. It will kill you just as efficiently as hydrogen sulfide.

Refinery distillation towers should be chemically cleaned for iron sulfide removal before the tower is opened for inspection. Note that iron sulfide is NOT water soluble. Citric acid is one solvent commonly used that can remove iron sulfide deposits.

Lessons Learned

If you learn nothing else from my book, absorb this:

- Go to your local hardware store.
- Purchase a first-class laser level light, which is accurate to a fraction of an inch (about $1,000).
- Check tray decks for being out-of-level. Being out-of-level a half inch is typically okay. Being out-of-level 2 inches will destroy distillation tray efficiency.
- Weir and downcomer bottom-edge levelness are also critical, but far easier to check, using a six inch ruler.

6

Distillation: Converting Wine to Brandy, plus Basic Calculations

Introduction

Mankind's introduction into the world of technology was initiated when a method was used to increase the alcohol content of beer (5% alcohol by fermentation) and wine (13% alcohol by fermentation) into alcoholic beverages of 80 proof (40% alcohol) or 120 proof (60% alcohol). The process was gradual until the invention of the partial condenser in Persia (now Iran) about 900 years ago and the "patent still" in the British Isles in the mid-nineteenth century.

Purpose

The problem that had to be overcome was that while alcohol has a greater vapor pressure than water, water also is a relatively volatile component. The ratio of the vapor pressure of alcohol (i.e., ethanol—not methanol or isopropyl alcohol) divided by the vapor pressure of water is called their *relative volatility*. The larger the relative volatility of these dual key components, the easier the separation. This means that, as we distill more of a liquid from a still, more product is produced, but with an incrementally higher water content (or lower proof). That is the problem of converting wine into brandy or beer into whiskey.

"Hi! I'm Norm. May I ask your name?"

"Susan."

"Sue! Can I buy you a drink?"

"I'm perfectly capable of buying my own drinks!"

"Susan. Are you interested in how the brandy you just ordered is made?" I asked.

"What? What did you say your name is?"

"Norm! You see, Susan, wine is only 13% alcohol. Brandy is 40%—or 80 proof—alcohol. When you ferment grapes, you can only get 15% alcohol—or 30 proof. Then, the yeast is killed off by the toxic alcohol or all the available sugar is fermented."

Relative Volatility

"Sue, take a glass of water and a glass of rubbing alcohol. Set them in a warm place. You'll find that a lot of the alcohol evaporates, but only a little of the water. The reason is that alcohol has three times the vapor pressure of water. If you divide the vapor pressure of the alcohol by the vapor pressure of water, that's called relative volatility:

- Relative Volatility $= \frac{VP_L}{VP_H}$

where:

- VP_L = Vapor pressure of light component
- VP_H = Vapor pressure of heavy component.

So, let's say I have a still, heated by a wood fire (Fig. 6–1). As I start to vaporize the wine in the still, the more volatile alcohol will preferentially vaporize. But the less volatile water will also vaporize, to a lesser extent. The rate of each component's vaporization is proportional to two factors:

1. Its vapor pressure
2. Its concentration

"Let's say that the vapor pressure of the alcohol is three times as high as the vapor pressure of the water, then the—"

"Norm, then the vapor will have three times as much alcohol as water," Sue surmised.

"No, Sue! The concentration of the alcohol and water in the vapor phase is also proportional to the concentration of each component in the liquid phase. The greater the concentration in the liquid phase, the greater the concentration in the vapor phase.

"By concentration, I don't mean wt. %, not the volume %, but the mole %. You've got to convert from wt. % to mole % by:

- Mole % proportional to (wt.%) ÷ (M.W.) where:

M.W. is the molecular weight of each component. The following are some common molecular weights:

- Ethanol = 46
- Air = 29
- Water = 18
- Carbon dioxide = 44
- Butane = 58
- Gasoline (typical) = 100
- Jet fuel = 150

Solubility of Phases

"The concentration of the alcohol and the water in the vapor from the still is proportional to the:

- (Mole % in the Liquid) • (Vapor Pressure),

assuming that the different components are completely soluble in each other."

"But, Norm, water and alcohol are always completely soluble, regardless of the concentration of either component."

"That's true, Sue. But if we tried to mix gasoline and water, we would form two separate phases: a hydrocarbon phase and an aqueous phase. And each phase would exert its own separate vapor pressure."

Batch Distillation

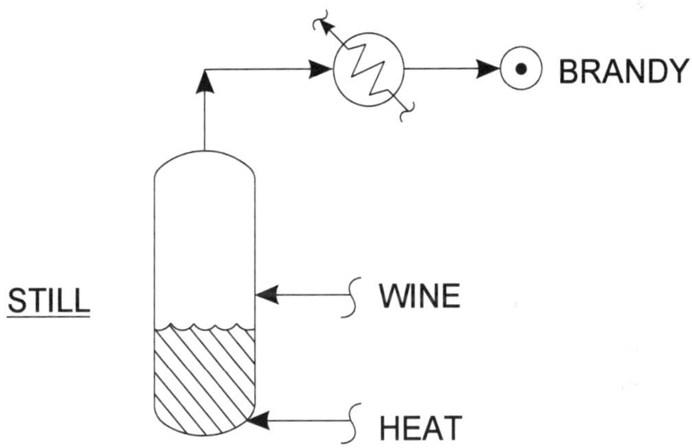

Fig. 6–1. A single-stage still

"So let's say I have a 10-gallon container of wine, which is 15 wt. % alcohol. And I start to evaporate it, and then I condense the evolved vapors. At first, the vapors I condense will be very rich in alcohol. I can calculate the % alcohol in the vapor in this way:

1. Convert the volume % of each component to wt. %, based on the specific gravity or density of each component.
2. Convert the wt. % to mole % (dividing by the molecular weight of each component).
3. Multiply the mole % by the vapor pressure of each component. This will tell me the partial pressure of each component (available in the API Data Book).
4. Add the partial pressures of the alcohol and water together. If our still is at sea level, this will equal 14.7 psia—atmospheric pressure.

5. Divide the partial pressure of the alcohol by the total pressure. That will equal the mole % of alcohol in our brandy."

"But Norm, why will the sum of the partial pressures exactly equal atmospheric pressure?"

"Because, Sue, the liquid in the still is boiling."

Adding Distillation Stages

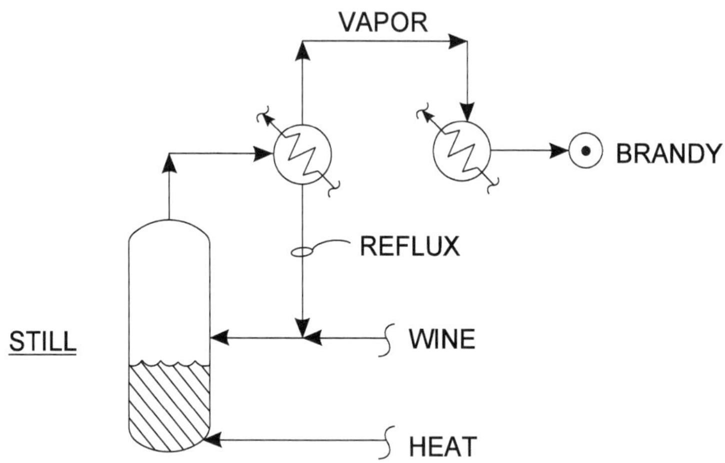

Fig. 6–2. Two-stage still

"Sue, the first vapors that evaporate from the still will be rich in alcohol. But then, as you continue to evaporate more and more of the liquid, and condense more and more of the vapors, the concentration of alcohol in the brandy will gradually diminish.

"It's a balance between product purity (percent alcohol in the brandy) and the percent of product recovery (the percent of the liquid in the still that was recovered as our brandy product).

"This problem—the balance between product recovery and keeping our product to a targeted specification—is what chemical engineering is all about. About 900 years ago in Persia (now Iran), one of my ancestors hit upon the solution to this critical ancient problem."

"Your ancestor?"

"Yes, Sue. Abu Ben Malik, my third-great-uncle. He added a partial condenser to the evolved vapors from the still. Malik condensed half of the vapors from the still. Of course, the water vapors would condense more readily than the more volatile alcohol vapors. So that amount of vapor that had not condensed in the first-stage condenser would now be condensed in a second condenser, to make alcohol-rich brandy.

"And the liquid, rich in water, from the first-stage condenser would be drained back into the still. Then, more heat would be provided to the still to revaporize, or reboil, the liquid drained out of the first-stage condenser, to recover more alcohol."

"And, Norman, what did your Uncle Malik call that liquid from the condenser?"

"Reflux!"

Mathematical Treatment

"Norm, could you actually detail what you have just said by showing me the relevant process calculations?" asked Sue.

1. Convert wt. % alcohol and water to mole %:
   - M.W. H_2O = 18
   - M.W. Alcohol = 46 (ethanol)
2. Container of wine has:
   - 72 lbs of water + 23 lbs of alcohol (i.e., 24 wt. % of alcohol)
3. 72 lbs water ÷ 18 = 4.0 moles water
   23 lbs alcohol ÷ 46 = 0.5 moles alcohol
4. Total moles in container = 4.0 + 0.5 = 4.5
5. Mole % alcohol = 0.5 ÷ 4.5 = 11%
6. Mole % water = 4.0 ÷ 4.5 = 89%
7. Vapor pressure of alcohol at 200°F = 53 psia
8. Vapor pressure of water at 200°F = 10 psia (The 200°F is the temperature of the wine boiling in the still.)
9. Partial pressure of alcohol = 11% • 53 psia = 5.8 psia
10. Partial pressure of water = 89% • 10 psia = 8.9 psia
11. Total pressure in still = 5.8 + 8.9 = 14.7 psia (i.e., atmospheric pressure at sea level)
12. % alcohol in vapor = 5.8 psia ÷ 14.7 psia = 40 mole % alcohol
13. To convert mole % to wt.:
    - For alcohol = 40 mole % • 46 M.W. = 18.4 lbs
    - For water = 60 mole % • 18 M.W. = 10.4 lbs
14. Wt. % alcohol in vapor initially leaving the still will then be:
    - (18.4) ÷ (18.4 + 10.4) = **63%.**

"Sue, to summarize, I started with 24 wt. % and increased this concentration by a factor of about 2.8, in what (as shown in Fig. 6–2) is a single-stage distillation column, plus a reboiler and a final condenser. The single stage of fractionation is the partial condenser. The reboiler is also a fractionation stage, but by convention, that is not included when we count the fractionation stages in a distillation column."

Lessons Learned

The principal lesson to be learned from this chapter is how to calculate the relative volatility of the key components and how to overcome a low-relative volatility with two factors: reflux (and reboiler heat) and distillation separation stages (or partial condensers). Distillation technology has not changed since the invention of the patent still almost 200 years ago, and it never will.

7

Distillation Tray Efficiency

Introduction

The petrochemical and refinery industry uses the following types of tray decks:

- Sieve Trays—½" or ¾" holes.
- Movable Valve Caps—These caps (1") flutter up and down with the vapor flow.
- Fixed Valve Caps—The cap itself is stationary. Escaping vapor flows preferentially toward the tray's outlet weirs.
- Bubble Caps—With an upside down tea cup on top of a short chimney.

Purpose

Sieve trays are extensively subject to tray leakage. Movable valve caps stick to the tray deck. Bubble cap trays are difficult to clean. The conclusion is that new distillation towers ought to be fitted with "Fixed Valve Caps" with a space between the cap and the tray deck of about 15—17 mm (known as lift). Also, to keep trays straight and stiff, tray deck thickness needs to be 10 gauge—not the industry standard of 14 gauge (bigger gauge means thinner metal tray panels).

I recently ran a computer simulation on a distillation tower, using the observed reflux ratio and the reboiler duty. I put in the following tray details, together with the gas chromatograph (G.C.) composition of the feed, tower operating pressure, and the feed conditions:

- Feed point
- Tray spacing
- Weir height
- Type of valves
- Valve lift
- Downcomer clearance
- Downcomer width

- Number of passes
- Tower I.D.

The computer simulation, using the actual number of trays, indicated that, theoretically, the fractionation efficiency was fine, if I assumed 60% to 80% tray efficiency. The problem was that the actual observed fractionation efficiency was terrible.

Expanding Tray Capacity in Small I.D. Towers Without Retraying

The following steps will expand tower tray capacity by 10–15%:

1. Drill ½-inch holes at the edge of tray, through tray support ring.
2. Insert ¼-inch spacer between downpour plate and lead perforated plate. This utilizes part of the area under downcomer for bubble area.
3. Reduce weir height to match downcomer clearance. Make sure weir and downcomer edge are *very* level.
4. Use "swept-back" weirs. Weirs are bent back at their edge.
5. Use "full radius" downcomer outlets. Bottom edge of downcomer is rounded, rather than sharp-edged.
6. Make sure tray decks and weirs are level, clean, and installed correctly. Check reflux distributor, seal pan installation, and the feed distributor for proper clearances.

Low Tray Efficiency

I am in favor of tower radiation density scans to discriminate between tray deck flooding or tray deck dumping. The beauty of the scans is that they permit me to troubleshoot a poor tray efficiency problem from my air-conditioned office in New Orleans. In a few minutes, using a radiation scan, I can decide if the observed low tray efficiency my computer model has identified is due either to dumping or flooding. Not only can one immediately make this observation, but the actual trays so affected can be readily seen from the scan. A wonderful tool!

But…scans are quite expensive. Especially on a 68-tray tower, with 2-pass or 4-pass trays. Also, from the time the unit engineer decides he needs such a scan to the time the vendor's report is finally received, likely a month has passed. For me, any period over a single day is excessive. Anyway, the cost for such a radiation scan is often excessive.

The Problem with Calculating Tray Pressure Drops

Assume we are not dealing with old-style bubble-cap or tunnel-cap trays that cannot leak, since the height of the riser is greater than the weir height. The vast majority of distillation towers currently in use are perforated trays: sieve, grid, or valve.

The fractionator I am discussing had modern Gridcaps. For such a cap not to leak (assuming a weir height of 3 inches), the pressure drop of the vapor flowing through the caps would need to be 1½–2 inches of liquid:

- $DP = K \bullet \frac{D_V}{D_L} \bullet (V)^2$

where:

- DP = pressure drop of vapor, in inches of clear liquid
- D_V = vapor density
- D_L = liquid density
- V = vapor hole velocity, feet per second
- K = orifice coefficient, typically between 0.3 to 0.6, depending on particular cap used.

The above 1½–2 inches pressure drop required to prevent tray leakage includes an allowance for the crest height. On the other hand, if the calculated pressure drop of the vapor flowing through the caps was 5–6 inches of liquid (assuming a tray spacing of 24 inches), then the poor fractionation would be due to entrainment or vapor flooding.

Theory vs. Reality

In the larger-diameter towers I've inspected, the tray decks will be somewhat out-of-level, especially after many years of service. The effect of out-of-level trays is to promote vapor-liquid channeling. The liquid drains through the tray deck sags. The vapor rises through the higher portion of the tray's panels. Result: poor vapor-liquid contacting and low tray efficiency. If the out-of-levelness of the tray decks exceeds the pressure drop of the vapor flowing through the tray's caps, then tray fractionation efficiency will be severely retarded.

Determining Cause of Low Tray Fractionation Efficiency

One standard method to discriminate between tray dumping and flooding is to measure the tray Delta P. If the average tray pressure drop is too low (i.e., less than 0.1 psi per tray), then likely the problem is low tray efficiency due to tray deck dumping, or leaking, or weeping. If the average tray pressure drop is too high (i.e., more than 0.25 psi per tray), then likely the problem is low tray efficiency due to tray flooding or entrainment.

The concept of discriminating between causes of low tray fractionation efficiency based on a differential pressure does not work well in practice. The problem is when flooding is initiated at an intermediate tray, rather than at a lower (or bottom) tray deck.

Flooding Progresses Up a Tower

I once had a propane-propylene splitter that was in fully developed flood. Opening the tower top vent indicated the liquid hydrocarbon was being forced out of the top head. Yet, the tower pressure drop of this 60-tray tower was close to zero. The problem was fouling of the top tray. The top tray was totally flooded, but since flooding may only progress up a tower and never down a tower, only the top tray was flooded. The other, lower trays had a low-vapor flow and hence a very low-vapor pressure drop. I suppose that these trays were dumping.

Varying Feed Rates at Constant Reflux Ratio

If trays are flooded because their valve caps are sticking to the tray decks, I would chemically clean the tower using a Zyme solution. But if fractionation is poor, because the tray panels are sagging and badly out-of-level, I would then open the tower and replace the conventional 14-gauge tray panels with 10-gauge, back-to-back, thru-bolted tray panels.[I] Especially in larger-diameter towers, out-of-level tray panels are often a major problem. Note that the smaller the gauge, the thicker the tray deck.

To see if a tray is weeping or flooding, I will proceed as follows:

- Step 1—Increase the feed rate by 20%.
- Step 2—Keep the same reflux ratio by increasing the reflux rate.
- Step 3—Increase the reboiler duty to maintain a constant level in the reflux drum.

I. Lieberman, N.P. *Process Design for Reliable Operations*, 2nd edition, Gulf Publications.

I'm currently working for Carl, the technical manager of an 80,000-BSD refinery in the Midwest. Carl's problem is bad fractionation in a naphtha fractionator. The tower makes two products:

- Light naphtha—Feed to a pentane-hexane isomerization unit, that changes straight-chain molecules to branched chains. Too many heptanes will coke-up the catalyst.
- Heavy naphtha—Feed to a naphtha reformer. Benzene precursors will increase benzene in gasoline product to an unacceptable level.

Fractionation will now either get better or worse as I apply the above procedure. If it gets better, it means that fractionation efficiency was being degraded due to tray deck weeping. If the separation efficiency degrades at higher rates, the problem was entrainment or flooding. If it improves, then the problem was tray deck weeping.

Back-to-Back Trays

The edge of each tray panel is bent down by 90 inches. Then the vertical edges are thru-bolted together. This creates, in effect, a half of an I-beam support at the edge of each tray panel. This type of assembly greatly increases the rigidity of the tray panels, and reduces the tendency of the tray to sag and hence develop low tray efficiency, especially at reduced vapor flow rates.

Equilibrium Separation Stage

I was the first person in history to design a complex fractionator that was actually constructed using a digital computer program. By complex fractionator, I mean a distillation tower that had multiple side-stream product strippers and draws, multiple pumparounds to remove heat, and several feeds (both vapor and liquid).

The tower was the "B" delayed coker at the American Oil Refinery in Texas City. Designed in 1966 and commissioned in 1968, it's still in operation at the Marathon Refinery (as of 2021). I visited it just last year. You can see it from Farm-to-Market Road 1764.

I had to create my own computer program in those early days of digital technology. Each theoretical stage (or tray) I represented by one *equilibrium separation stage*. This consisted of mixing liquid from the stage above with vapor leaving the stage below, at 100% mixing efficiency—then, separating the resulting mixture of phases, with 100% separation efficiency. Both the enthalpy of the vapor and liquid, and the molal components of vapor and liquid, I assumed were perfectly mixed.

For my design, I increased the actual number of trays by 30% above the number of the theoretical separation stages. I helped start up the tower and had the unpleasant experience of observing a design error. I had forgotten about avoiding the water dew point at the top of the tower. I discovered this when I drew a sample of the top side-cut (heavy naphtha) and found a few percent of water in the sample bottle. I'm smarter now.

Lessons Learned

Tray efficiency is a function of:

1. Proper tray vapor-liquid loading process calculations.
2. Proper specification of the tray process and mechanical details.
3. Tower internal inspection prior to re-streaming the distillation tower.

8

Steam Stripping Products

Introduction

Steam stripping refinery gas oil to maximize virgin diesel production is the most important aspect of technology to maximize refinery profitability. Stripping efficiency is a function of the number of stripping trays, the amount of stripping steam, dryness of the steam.

Purpose

The most important aspect of hydrocarbon stripping is the moisture content of the stripping steam. Steam with 2% moisture is marginally okay. But steam with 10% moisture greatly reduces the efficiency of the steam.

Stripping tray efficiency is also a major factor. The mass flow of vapor through a steam stripper with five to ten trays increases by a factor as much as ten from the bottom tray to the top tray. Designing trays for such a large change in vapor flow is a major challenge.

Wet stripping steam is bad because the water in the steam hits the hot oil and evaporates. It takes heat to evaporate water. This heat comes from the oil being stripped. The water subcools the oil, and then the lighter hydrocarbons in the oil don't evaporate, if the oil is cooled by the wet steam.

What Does Wet Steam Look Like?

Dry steam is invisible. If it blows white out of a bleeder, that white stuff is water. If it is "dripping white," that's really wet steam. It could be that the steam trap is defective. Or, if the steam is being generated in a nearby waste heat boiler, it could be the level is set too high. Or maybe there is a knock-out drum on the steam supply that isn't draining properly.

Once, I improved a stripper's efficiency by wrapping the steam supply line with rags to dry the steam out and by installing a new steam trap. This allowed the refinery (Coastal—Eagle Point, New Jersey) to meet their diesel oil flash specification one cold winter night.

Effect of Pressure

The idea of stripping with steam is to blow away the lighter hydrocarbons, such as gasoline, away from the heavier hydrocarbons, such as diesel. So, the lower the pressure in the steam stripper, the better the stripping efficiency. Check the pressure at the top of the stripper as you increase the steam flow. If the pressure goes up one psi, that's normal and okay. But, if the pressure goes up by 3 psi, then you're using too much steam and need to back off on the stripping steam supply valve. An increase of 5 psi on a diesel oil stripper has the same effect as subcooling the diesel product by approximately 10°F.

Effect of Temperature

The heat to vaporize the lighter components out of the stripper bottoms product doesn't come from the steam. It comes from the oil itself. That means the temperature of the oil is used to vaporize the gasoline out of the diesel.

Let's say you're not using any steam in the stripper. Then, you would expect the temperature of the feed to the stripper would be just a bit hotter (maybe 2°F) than the liquid leaving the bottom of the stripper.

Now you turn on the steam. Assume the steam is dry. Then, as the steam strips the lighter gasoline out of the diesel, the diesel will cool off by about 25°F (15°C). If that doesn't happen, then the stripper is not working properly.

Feeding the Stripper

How does the unstripped diesel get into the side-stream stripper (Fig. 8–1)? It isn't pumped. It flows by gravity out of the fractionator into the stripper. That is, the draw-off nozzle from the main tower is about five to ten meters above the inlet to the stripper. This draw-off flow passes through a level control valve. The level I'm referring to is the level in the bottom of the stripper.

The flow from the bottom of the strippers is on flow control. So, as you increase the flow from the bottom of the stripper, to make more diesel, the stripper's bottom level goes down, which tells the feed inlet control valve to open.

Note that an excessive pressure drop in the stripper overhead vapor line may stop the flow into the stripper. This is a very common problem and forces the operator to reduce the stripping steam flow rate.

What Are We Trying to Accomplish with Steam Stripping?

A diesel steam stripper is trying to meet a "flash point" spec. That is, the temperature at which the diesel will ignite when exposed to a flame. The flash point spec for diesel is typically 150°F. A flash point less than 150°F indicates that the diesel is not being stripped hard enough. Meaning, there is too much gasoline in the diesel product. You will need to turn up the stripping steam valve.

Wait 30 minutes, and then get another sample. If more steam doesn't help, it could be the steam is wet or the stripper pressure increases by using too much steam. Higher pressure suppresses vaporization of the lighter components in the stripper's feed.

What's Inside the Stripper?

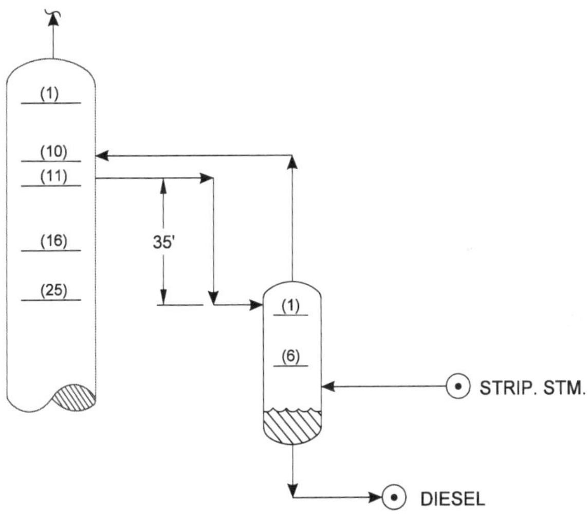

Fig. 8–1. Side-stream stripper

Inside the stripper, there are trays (Fig. 8–1). There are typically six perforated plates. The tray is a way to mix the steam and diesel together. More trays should mean greater stripping efficiency.

What we don't want is for the liquid level in the bottom of the tower to get too high. This can damage the trays in just a few seconds. You should be able to see the yellow diesel in the stripper's bottom gauge glass. You'll have to learn how to operate the "Jurgenson ball check valve" at the top and bottom of the gauge glass. If the liquid level rises above the steam inlet nozzle—even for a few seconds—you will blow the trays off their support rings.

Packed Strippers

In many refineries, the trayed stripper has been retrofitted with packing—either rings or a bed of structured packing. Most of these retrofits do not work. My response is always to replace the packing with "grid-type" trays. Each tray has a number of caps in proportion to the vapor flow.

Tray Hole Areas

The vapor flow greatly increases from the bottom tray to the top tray. Therefore, the top tray will require roughly three times the hole area of the bottom tray. The bottom tray vapor flow is only steam. The top tray vapor flow is the steam plus all the hydrocarbon stripped out of the feed.

Drawing Samples

Diesel oil autoignites at about 400°F. So, if you get a really hot sample and spill it, you may cause a fire, even if no flame is present. Get samples from a sample cooler—not hot from the discharge of a pump, through a bleeder valve. The heavier the product, the lower the autoignition temperature and the greater the danger of sampling. If there is no sample cooler, switch pumps. Shut off the seal flush to the pump which had been running, and wait an hour. Then get your sample from the idled pump.

The Vacuum Tower

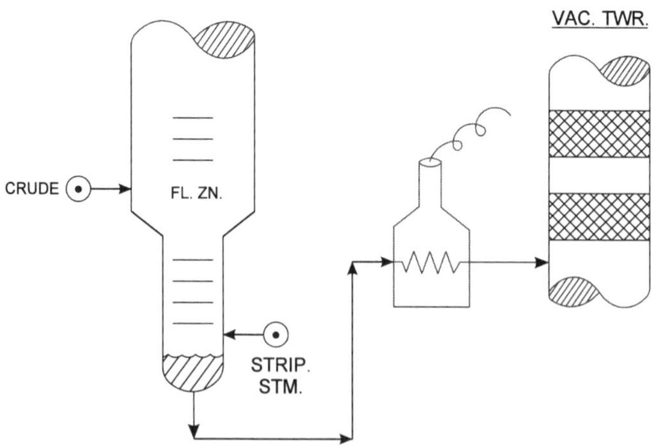

Fig. 8–2. Steam Stripping Vacuum Tower Feed

The crude tower feeds the downstream vacuum tower (Fig. 8–2). The better the steam stripping of the vacuum tower feed in the bottom of the crude tower is, the better the vacuum developed in the downstream vacuum tower, and the more efficiently the vacuum tower will work. If you use too much stripping steam, and that increases the crude tower flash-zone pressure, that may also reduce the operating efficiency of the vacuum tower. Half the time, when I increase a crude tower bottom's stripping steam, the result is loss of vacuum. It's necessary to experiment.

Lessons Learned

In one of my first jobs on a steam stripper that wasn't working, it was because the stripper insulation had fallen off and the stripping steam was wet. So, I shut off the steam. The stripper pressure went down, its temperature went up, and the diesel's flash increased by a few degrees.

Steam stripping is complicated. You have to understand the fundamentals and pay careful attention to what you're doing as you adjust steam flow. Here are a few key points:

- Keep stripping steam as dry as possible.
- Insulate the stripper feed line and the stripper itself.
- The top tray of a steam stripper typically has three times the hole area as does the bottom tray of the stripper.
- Distribute the stripping steam flow through a steam dispersion piping grid below the tower's bottom tray.

9

Draft in Fired Heaters

Introduction

Most of our heaters are natural draft. The less common induced-draft and forced-draft heaters are not discussed in this chapter. Lack of adequate draft will cause a positive pressure to develop beneath the lower row of tubes in the heater convective section. This may be a dangerous situation (due to SO_2 emissions) and undermines the structural integrity of the convective section.

Purpose

A slight negative pressure or a positive draft of 0.1 inch H_2O must be maintained immediately below the bottom row of convective tubes. This is not only my idea—it is your company's safety requirement and is not subject to your interpretation. If the draft is too big, excessive amounts of cold tramp air are drawn into the convective section. This is not hazardous, but it wastes energy by reduction of the heat pick-up in the convective section. A draft of 0.2–0.3 inches is excessive, below the convective tube section.

"Señor Norm. Por favor."

"Que pasa, mi amigo Pedro?"

"Please, Sir. Look over this letter to Foster Wheeler in New York. Is it all correct? All in proper English?"

I glanced through the letter. It described a heater flame impingement limitation at the Ecopetrol Cartagena Refinery, where I was working in Colombia. The letter suggested that the natural draft heater was draft limited. It authorized Foster Wheeler Corporation in New York City to calculate, using the appropriate computer simulation software, available draft in the crude heater, for a fee not to exceed $10,000 (USD).

"Pedro, $10,000! Mucho dinero! Are you really draft limited? How high is the heater stack?" I asked.

"Señor Norm. The stack is very tall. Maybe 50 meters. When we open up the burners' air registers, flue gas blows out of the site ports just beneath the bottom row of convective tubes. Muy peligro! We are burning 3% sulfur fuel oil."

"Pedro! You can kill someone with SO_2 blowing out of the site ports. You surely do have a positive pressure at the top of the radiant section. I presume the stack damper is wide open?"

"Si! Si! Señor Norm, that is why we have written to Foster Wheeler, the world expert on fired heater design, for help in the calculation of the stack's draft," Pedro sighed deeply. "Muy, muy importante!"

"Pedro. Let's drive out to the plant. Best to look the problem over in the field. You drive, and I'll tell you a story."

Draft: A Function of Density Difference

I have never progressed much from childhood. The tests I run in process plants are the same as my experiments as a child. Often, my mother would scream at me through my bedroom door,

"Norman! You have to leave for school at 7:30!"

"Mom, I can't! I'm doing an experiment."

One of my experiments was a balloon lifting test. I had a large, red, helium-filled balloon. I had lead weights that I used for fishing. The experiment was to find how many of the small lead weights the balloon would lift, so that it would be suspended.

When I became older, I expanded my experiment to include calculations. I had learned in school that the molecular weight of helium was 4 pounds and that the molecular weight of air was 29 pounds. I had also learned that the density of air is 0.076 lbs/ft^3. This is the air density at sea level and at 70°F. The density of the helium in my balloon was lower, because, as noted, the molecular weight of helium is only 4, and air is 29.

- Density of Helium = 4/29 • (0.076) = 0.010 lbs/ft^3

The density difference between the helium inside the balloon and the air outside the balloon could be calculated this way:

- 0.076 – 0.010 = 0.066 lbs/ft^3

My balloon was round, so I could calculate its volume, which was 3 cubic feet. Therefore, I calculated the lifting force of my balloon:

- (0.066) • 3 = 0.20 pounds

As there are 16 ounces in a pound, I could figure the **LIFT** on the balloon:

- 0.20 x 16 = 3.2 ounces

Using a postal scale, I cut out paper weighing 3.2 ounces—which, to my disappointment, the balloon could not lift. I had forgotten about the weight of the balloon itself. Actually, the balloon could only lift about 2½ ounces, so I guess the balloon itself weighed 0.7 ounces:

- 3.2 – 2.5 = 0.7 ounces

Effect of Temperature on Lift

Helium is expensive, and hydrogen is dangerous due to its combustibility. So, if you want to go on a balloon ride, the balloon is going to be lifted by hot air. A balloon 40 feet in diameter will have this volume:

- $4/3 \cdot \pi \cdot R^3 = 32{,}000$ cubic feet

Ambient temperature is 70°F. Air in the balloon has been heated to 300°F. The elevated balloon temperature makes the molecules of air push away from each other and thus, the air becomes less dense. The air is less dense in proportion to its temperature. But not in proportion to °F, but in proportion to the air temperature in °R (Rankine). The following formula can be used to convert from °F to °R:

- °R = °F + 460

In °R, the ambient air temperature is then:

- 70° + 460° = 530°R.

In °R, the 300°F balloon temperature is then:

- 300° + 460° = 760°R.

If the density of air at 70°F is 0.076 lbs/ft^3, then the density of the 300°F (760°R) hot air inside the giant balloon can be calculated:

- $0.076 \cdot 530/760 = 0.053$ lbs/ft^3

This then gives us the density difference between the hot, 300°F air inside the balloon and the cool 70°F air outside of the balloon:

- 0.076 – 0.053 = 0.023 lbs/ft^3

Since the balloon's volume is 32,000 ft^3, the balloon can lift the following weight:

- (0.023 lbs/ft^3) • (32,000 ft^3) = 740 lbs

In other words, a balloon and basket weighing 250 pounds, plus three people weighing 160 pounds each, could be lifted.

Stack Draft

A tall stack can develop a real strong sucking force. Let's say the stack's draft is 2 inches of water. That means the pressure outside of the stack is 2 inches higher than the pressure inside the stack. Let's say there is a metal plate 2 feet by 2 feet that I've been told to pull off at the bottom of the stack. How much force, in pounds, do I have to pull on the plate?

We need to figure the area of the plate:

- 2 ft. x 2 ft. = 4 ft^2

or:

- 4 ft^2 x 144 in^2/ft^2 = 576 in^2

Over the surface of this plate, there is a force of atmospheric pressure equal to 2 inches of water pushing on the plate:

- (576 in^2) • (2 in) = 1152 in^3

A cubic inch of water weighs 0.58 ounces. So, I would have to apply the following pulling force on the plate:

- 1152 in^3 • 0.58 = 664 ounces

or:

- 664 ounces ÷ 16 ounces per pound = 42 lbs

I actually had this problem in the Texas City Amoco Refinery in 1980. In this instance, I had to pull off a cover to a manway on a reactor that was 150 feet high and full of hot flue gas from a burning sulfur residue. I was easily able to do this.

Effect of Stack Height

Draft in a stack is not caused by combustion. Also, the molecular weight of the flue gas is rather close to air, and this effect is neglected. Wind blowing across the top of the stack certainly does contribute to draft. But, since it is not always windy, the effect of the wind is ignored when calculating draft.

You may have seen on certain stacks that the top end of the stack is necked-down (Fig. 9–1). This is done to slightly increase the draft. It's called the "Venturi effect."

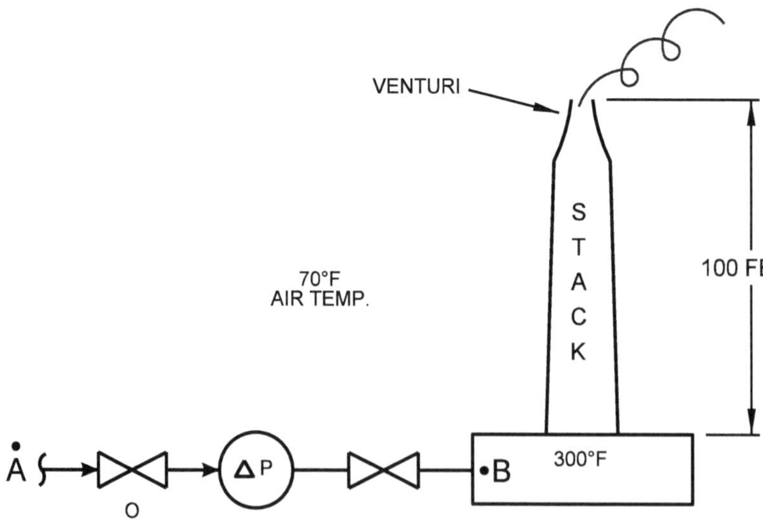

Fig. 9–1. DRAFT = 0.45" H_2O)

Here are the major things that contribute to draft in a stack:

- The height of the stack
- The temperature of the flue gas
- The ambient air temperature

Let's say a stack is 100 feet high. The ambient air temperature is 70°F, and the temperature of the gas inside the stack is 300°F. The molecular weight of the gas inside the stack is pretty close to air. If you refer back to my hot air balloon calculation, you see the density difference between the cool 70°F ambient air and the hot air (300°F) inside the balloon:

- 0.023 lbs/ft^3

This density difference applies to the entire 100-ft. height of the stack. It's kind of like the cold, denser air pushing up the lighter, hotter balloon. Meaning, the 100-ft. column of colder air is pushing up the 100-ft. column of hotter air inside the stack. This pressure difference is called draft. It's a result of density difference multiplied by the 100-ft. height of the stack:

- Draft = (0.023 lbs/ft^3) • (100 ft.) = 2.3 lbs/ft^2

Of course, anyone who has ever worked on a process heater knows draft is never expressed in "lbs/ft^2," but in "inches of water." So let's convert lbs/ft^2 to inches of water, in two steps:

- Step 1—Divide the stack draft by 144 square inches per square foot:
  - Draft = (2.3 lbs/ft^2) ÷ (144) = 0.016 lbs/in^2

The term "lbs/in^2" is normally referred to as pounds per square inch, or simply as psi. But how do you convert from psi to "inches of water"? If you fill a glass tube with water to a height of 28 inches, and then place a pressure gauge at the bottom of the tube, your gauge will read 1.0 psi (Fig. 9–2). This means that to convert from psi to inches of water, I multiply the psi by 28. So we can find the draft in our 100-foot-high stack full of 300°F stationary air:

- Step 2—Perform the following calculation:
  - Draft = (0.016 psi) • (28 inch H_2O/psi) = 0.45 inches of H_2O negative pressure.

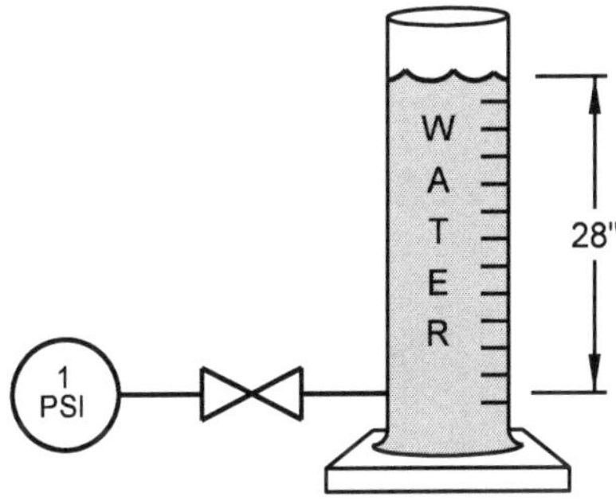

Fig. 9–2. Head pressure of 60°F water.

Don't We Need Fire to Create Draft?

What we require to create a draft in a stack is heat! Heat to warm up the air from 70°F to 300°F. If I heat up the air inside a stack with an electric heating coil, and I put a cap on top of the stack to prevent flow, I would still measure that the pressure inside the stack at the bottom of the stack would be 0.45 inches of water lower than the ambient air pressure outside of the stack.

As the gas inside the stack gets hotter and less dense, draft increases. As the air outside of the stack gets colder and more dense, draft also increases.

Because the following are small effects, I neglect three things:

- The frictional losses of the flue gases running up the stack. Stacks are made fat enough for this to be true (i.e., velocities less than 50 feet per second)
- The molecular weight difference between air and flue gas
- Wind blowing across the top of the stack

I have assumed in the above discussion that the pressure at the top of the stack is ambient, 100 feet above the base of the stack.

Effect of Wind

I have observed that a 30-mile-per-hour (mph) gust of wind will cause an increase of draft of 0.1 inches of water. This will increase the combustion airflow into a fired heater. If you're running a heater with 2% excess O_2 in the stack, the excess O_2 will suddenly increase to about 4%.

Or, if you're running a heater at 2% excess O_2, and the 30-mph wind stops, you're going to be running the heater in an air-deficient mode, and explosive gases may accumulate in the convective section! Think about the consequences, if the fuel gas is on automatic temperature control.

The Venturi Effect

I was driving down a gravel road in southern India, and I drove past a small plant boiling down sugarcane juices to produce molasses. I noticed at the top of the 20-inch-diameter brick boiler stack that the stack diameter swedged down to perhaps 6 or 8 inches.

I reasoned that this reduction in the cross-sectional area of the stack would increase the velocity of the flue gases exiting the top of the boiler's stack. Such an increase in the velocity of the flue gas would decrease pressure at the top of the stack. This should result in a slight increase in draft in the boiler at the base of the stack, to help draw in more combustion air.

I stopped to ask the two guys working around the sugarcane juice boiler why they thought the top of their stack was swedged down. I suggested to them that this was an application of the "Venturi effect."

And they responded, "Sir! This narrowing of the top of the boiler's stack is not done to honor Venturi. It is done because narrowing the top section of our stack helps the boiler to work harder."

The Fix

Returning to my story in Colombia, Pedro and I drove for 30 minutes, past banana groves and hillsides covered by withered corn stalks. Onsite, the crude heater showed many signs of operating in a draft limited situation (Fig. 9–3):

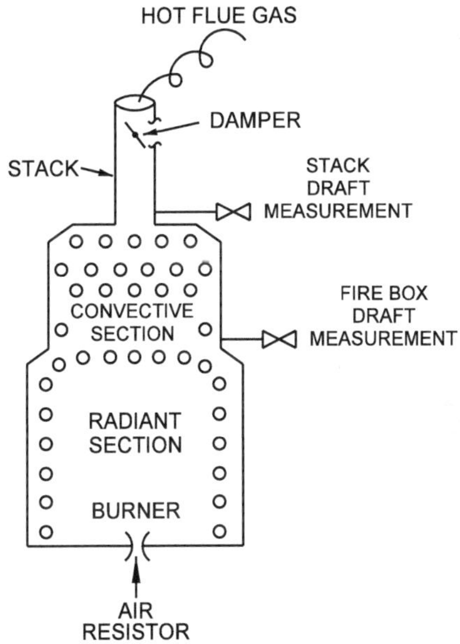

Fig. 9–3. Components of a natural draft fired heater.

- The burner air registers were mostly closed.
- The yellow flames were licking up against the radiant section wall tubes.
- Wisps of white (i.e., SO_2) flue gas were escaping from leaks just below the convective section.
- Increasing fuel oil flow caused the heater outlet temperature to slip down—a sure sign of lack of combustion air.

I looked up at the stack damper position indicator arrow. The arrow pointed straight-up, indicating that the damper was wide open. I then proceeded as follows:

- Step 1—Using a bottle of blue-colored water and 4 feet of clear plastic tubing, I measured the draft above the convective section tube bank. It was about 0.3 inches of water. Meaning, that's how high the blue water was sucked up into the tubing by the slight vacuum at the base of the stack.
- Step 2—I calculated the available draft that should have been generated by the stack, based on the flue gas temperature and the stack height, as I described above. It was 0.9 inches of water.
- Step 3—I concluded that the draft loss, or pressure drop of the flue gas, through the damper was (0.9 – 0.3) 0.6 inches of water.
- But how could a wide-open stack damper be causing a pressure drop equivalent to 65% of the stack height? So I then proceeded as follows:
- Step 4—While observing the blue water level in the plastic tube, Pedro slowly moved the stack damper indicator arrow, until the arrow moved from a vertical position (straight-up) to an inclined, 45° position. And the draft, which had been reading a "rock steady" 0.3 inches of water....did **NOT** change!
- Step 5—I had a crane with a basket brought up next to the heater's stack. Then, I had Pedro ascend to the stack's damper shaft with a 30-inch pipe wrench. He twisted open the damper until the blue water in the plastic tube rose to 0.9 inches, at which point I concluded that the stack damper was actually wide open. The problem was that the stack damper chain operator was only moving the damper position indicator arrow, but not the damper itself.

With this change, the crude heater flames withdrew from the walls, and the burner air registers could be opened without blowing SO_2-rich flue gas out of the fire-box roof.

The Final Judgment

When I told my parents about this adventure, my dad said, "Good work, son. You used your wits to solve an important problem."

But my mom said, "So, did those Colombians give you the $10,000?"

And I said, "Mom! The reward is not about money. The satisfaction comes from the job itself."

And Mom said, "Thank goodness your grandmother is not alive to hear that."

Lessons Learned

- Opening either the primary or secondary burner air registers will reduce draft.
- Opening the stack damper will increase draft.
- Increasing the stack temperature will increase draft.
- Colder ambient temperatures will increase draft.
- Higher firing rates may or may not increase draft, depending on the effect of stack temperature.
- Tramp air leaks in the convective section will definitely reduce draft.
- Increase in excess air will also always reduce draft.
- High wind velocity increases draft.
- Fouled convective section fins decrease draft.

10

Hazards in Fired Heater Operations

Introduction

This is the most important chapter in this text. Many operators are killed by a heater getting caught up in a positive feedback loop, due to lack of combustion air. I lost two friends at the Good Hope Refinery in 1982, by failure to understand the importance of not falling below the "Point of Absolute Combustion," which is the main subject of this chapter. Also of great importance is the correct use of "snuffing steam", when a fired heater must be taken offline in the event of a tube rupture.

Purpose

When trying to save fuel, we often reduce excess air too much. This creates the positive feedback loop explained in this chapter. Positive feedback is good in relationships. But it's really bad in a process control loop, when controlling heater fuel flow based on the heater's process outlet temperature. Trying to optimize the combustion air rate, to save on heater fuel, makes the problem much more common.

The correct use of snuffing steam is not related to the above, but is a critical safety item when a process tube is ruptured in hydrocarbon service.

"Can you explain to us why it is that sometimes we cannot control this heater 'on auto'? I've had all the instruments checked several times, and they're all okay," asked Paul, the control board operator.

I had walked down to the control room with the process engineer. The shift supervisor, hearing the discussion, had joined us by the visbreaker H-4 control panel. The visbreaker heater was running along smoothly on automatic control as we talked.

"Tell me what happens when you lose control of the heater on auto," I responded.

"It happens just like this. The heater can be firing normally, but sometimes when the heater outlet temperature drops, although the fuel gas control valve on

H-4 will open on auto as expected, the heater outlet temperature continues to fall. Each time that this happens I have to go back to manual control," Paul explained. "Is it the fuel gas control system that is faulty, or should we open up the burner's air registers when this happens?"

"It's not that simple. It's probably not a malfunction with instrumentation or the fuel gas control loop. I certainly would not recommend suddenly opening the air registers," I said. "Do you remember Gene, the process engineer? Here's how I explained this situation to her."

Gene was really excited. She had just returned from a fired heater energy conservation seminar in Houston. Full of enthusiasm, she burst into my office.

"Norm! I'm going to save the refinery a huge amount of money by reducing excess oxygen in our heaters. They taught us at the seminar to target for 3% excess O_2 in the heater stack (Fig. 10–1)."

"Gene, you need to be careful."

"Okay, Norm! I'll get the stack oxygen analyzer recalibrated. I'll tell the operators on number 3 Pipestill to reduce the flow of combustion air real slowly and carefully to the heater. We're going to save millions and millions of BTUs! I'll start today."

"Gene, I don't want to discourage your sense of adventure. But could you listen to a short story first?"

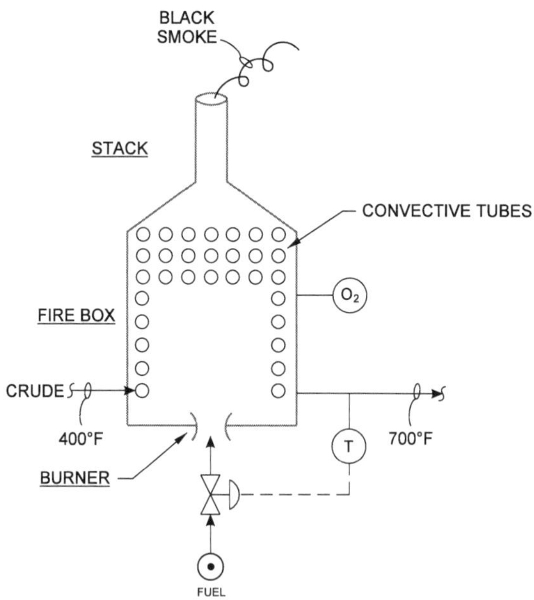

Fig. 10–1. O_2 Analyzer located correctly below bottom row of convective tubes avoids effects of tramp air leaks

Dangers of Positive Feedback Loop in Fired Heaters

Let's say I'm operating a refinery fired heater on automatic temperature control (Temperature Record & Control; TRC). This means that the heater's outlet temperature is automatically adjusting the fuel rate, to keep the heater process outlet temperature constant. Let's say this target temperature is 300°C. The combustion airflow is manually regulated by adjusting the secondary air register openings, on the fuel burners. But, because the outside operator is not available at this moment, the flow of combustion air is not being adjusted; rather, it is constant.

As the fuel rate increases automatically, because the heater's process outlet temperature is below 300°C, the amount of O_2 in the stack's flue gas will go down. Which is fine, as long as it doesn't get too low.

If the fuel mixes really well with the combustion air, then having as little as 0.5% O_2 in the stack flue gas is okay. If the fuel mixes really poorly with the combustion air, then having as much as 5% O_2 in the stack flue gas may be too little.

But what do I mean by too little oxygen in the flue gas?

What I mean is that, if you reduce the flow of air a little and keep the fuel rate the same, the heat of combustion of the fuel will go down. Instead of all the carbon in the fuel burning to CO_2, some of the fuel's carbon will burn into CO (carbon monoxide). Burning carbon into CO releases a lot less heat than burning carbon into CO_2.

How well the fuel and air mix depends on burner efficiency, which is very variable, depending on the design and condition of a burner. Let's assume a burner efficiency is such that an O_2 of 2% in the stack flue gas is just about right. Meaning, that having 1.5% O_2 in the stack flue gas decreases the carbon burnt into CO_2 and increases carbon burnt into CO. Then the result of adding more fuel, without more air, causes the heat generated in the firebox to diminish.

This would be no big deal, if the fuel was adjusted manually. But if the fuel adjustment is on TRC, then the reduction in heat generation by adding more fuel will have the following results:

- The heater process outlet temperature will drop below 300°C.
- The fuel flow rate (but not the air rate) will increase.
- The increased fuel flow rate will drive the heater outlet process temperature lower still, because of increased CO in the stack flue gas.
- This causes the fuel flow rate to increase and increase, as the heater process outlet temperature slides further and further down.

The heater is now caught up in a **positive feedback loop**, which is not a good thing. With time, other adverse effects will occur:

- Black smoke will be emitted from the heater stack.
- The heater firebox will become hazy with long, licking, yellow, smoky flames.

- Fire will eventually race out of the top of the stack.
- The heater outlet temperature will continue to diminish, slipping down and down, while the fuel gas rate increases exponentially.
- If the outside operator now intervenes, by manually increasing airflow suddenly through the burners, the heater will blow up.

What the consul or panel board operator should have done is to first switch to the manual mode of control for the fuel flow rate. Then back down the fuel flow rate until the process heater outlet temperature starts to go down and, only then, is it safe for the outside operator to increase the combustion airflow rate manually through the burners' air registers.

The Danger of Variable Wind

Wind blowing across the top of a heater stack will increase draft and draw more combustion air into the firebox. If the wind suddenly diminishes, the flow of combustion air will momentarily drop. If the heater is running too close to its optimum combustion air rate, it will now drop briefly below this optimum air rate. The heater outlet temperature will then drop and, if the heater is operating on TRC, the following sequence will occur:

1. The fuel flow rate will increase.
2. The heater outlet temperature will decrease further.
3. The fuel flow rate will increase even more.

The heater is now caught up in a potentially dangerous positive feedback loop. Suddenly opening the burner air registers to break the loop can—and has—blown up the heater firebox.

The Point of Absolute Combustion

I call the air-to-fuel ratio that maximizes the heat generated or released by burning a pound of fuel the "Point of Absolute Combustion."

Decreasing the air-to-fuel ratio below this point reduces the temperature of the burning fuel. The amount of carbon monoxide, light alcohols, ketones, and other products of partial oxidation of fuel will increase exponentially. The operators will find their heater cannot be run on automatic temperature control and that the fuel rate has to be controlled manually. Certainly, operating a fired heater below the Point of Absolute Combustion also wastes energy.

If the fuel is a liquid, heavy industrial fuel oil (Bunker "C"), used on ships and power plants, dropping the air-to-fuel ratio below this point will cause the color of the flue gas emitted from the heater's stack to go black. If the fuel is a gas, the color of the flue gas from the stack, when operating below the Point of Absolute

Combustion, may turn pale yellow or slightly orange. If the fuel is hydrogen, the flue gas color will stay clear, even when combustion airflow is deficient.

The important concept to remember is that the Point of Absolute Combustion will vary with time depending on the air-fuel mixing efficiency of the burner. Typically, I have seen it vary from 0.5% to over 6% excess O_2 in the stack's flue gas.

"Gene, I will tell you two more stories, related by plant operators to illustrate this concept."

Effects of Air on Visbreaker Heater Outlet Temperature

I was teaching a seminar a few years ago. A young operator had a question.

"Mr. Norm. My visbreaker unit is real old. The heater is old, and the controls are old. We run the heater fuel on manual because the controls are so old. Last week, by accident, I turned on our third forced draft air blower."

"Okay, then what?" I queried.

"The heater outlet jumped from 680°F to 720°F, with the same fuel flow. I never moved the fuel flow. What do you make of that?"

"What did you do then, when you saw the heater outlet temperature jump up?" I asked.

"I shut the blower down, Mr. Norm!"

"You should have cut the fuel rate back and left that third blower on. You must have been running the heater in an air-deficient manner before you turned on the third air blower," I explained.

"But Mr. Norm. We had 3% excess O_2 on our stack gas analyzer with just two blowers on. Our target is 3% O_2 . We were given this target by the plant manager. With the three blowers on, our excess O_2 was over 6% in the stack! I'm really confused, Mr. Norm. Should I run at 3% or 6% oxygen in the stack?"

"What are you trying to accomplish? Save fuel or save air? You have to think about what you're doing. Most likely, you have a lot of tramp air leaks in the convective section."

"Well maybe the 6% would be better to save fuel. But my main objective is to get paid and keep my job. Our plant manager set the target at 3% in the stack. If we run the heater above that target, then it's time off without pay."

Effect of Air on Boiler Steam Generation Rate

Here is the story of an old utility plant operator, George.

"Mr. Lieberman. I've got a target of 2%–3% O_2 in the stack of my 650-psig steam boiler. I run at a constant fuel rate. Our VP has set that O_2 target himself!"

"George, what's your point?"

"Mr. Lieberman, the point is that, when I increase air to 4% O_2, steam production goes up at a constant fuel gas rate. That's my point!"

Location of O_2 Analyzer

"Gene, take a look at (Fig. 10–1). See that I've shown the oxygen analyzer located below the bottom row of convective tubes. If the O_2 analyzer is located in the stack, as many of these are (Fig. 10–2), then tramp air leaks in the convective section can cause the analyzer to read higher than the O_2 in the firebox, where combustion is actually taking place.

"If the leaks are really large, then closing the air registers at the burners can increase the O_2 reading in the stack. This happens because restricting the airflow through the registers increases draft, i.e., reduces the pressure in the firebox. This creates more tramp airflow into the convective section and influences the O_2 analyzer, if the analyzer is located in the stack.

"Don't worry about this, because on our No.3 pipe still, the analyzer is located properly as shown in (Fig. 10–1)."

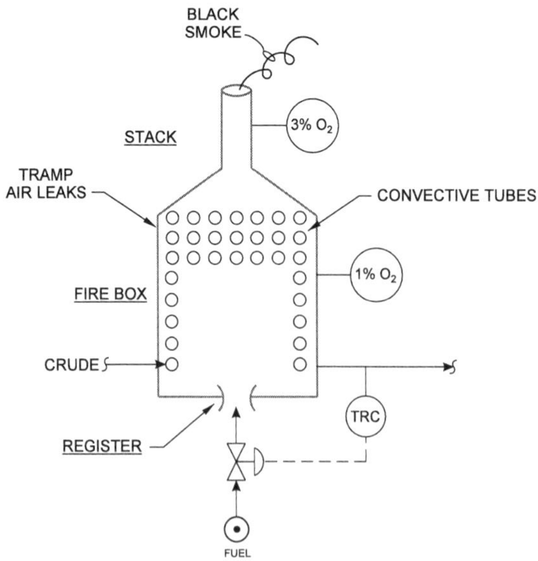

Fig. 10–2. Effect of tramp air leaks

Summary

Improving air-fuel mixing efficiency reduces the amount of excess air needed to maximize heat release from a given amount of fuel. If a heater is operated on automatic temperature control, which is typically the case, then falling below this optimum amount of combustion air will cause the fuel rate to increase, while the heater outlet temperature declines. This is called by plant operators "flooding the firebox with fuel" or "stalling the firebox."

A sudden increase to the proper amount of airflow, by opening the burner secondary air registers and / or rapidly opening the stack damper, can—and has—caused process heaters to explode, with resulting fatalities.

A rather similar incident happened to me personally in 1980, during the strike at a refinery in Texas City. I had inadvertently closed the air register of the sulfur plant incinerator. Seeing that the incinerator temperature was rapidly dropping, I realized my mistake. And, without thinking, rapidly opened the air register. The resulting "thump" knocked me off my feet.

"Of course, Gene, that was a long time ago."

"You know, Norman, you're right! I've got a good idea. What I'll do is put the fuel on manual first. Then I'll try increasing and decreasing the combustion airflow through the burner air registers. I'll find out what excess O_2 in the stack maximizes the heater outlet temperature, with a fixed amount of fuel. Then I'll use this as a target for determining the Point of Absolute Combustion. What do you think, Norm?"

"Excellent, Gene! But always run somewhat above the target O_2 and re-check the target O_2 periodically. Remember that the optimum O_2 is a variable. It depends on all these factors that we are discussing. What you are planning is pretty much the same way as I saw the operators controlling the Murphy Oil Crude heater in New Orleans a few years back. I recall that it was working quite well. The Point of Absolute Combustion for that heater was an excellent 0.56% O_2 in the firebox. The lowest I have ever experienced."

Conclusion

Back in the central control room at Good Hope, I could see that Paul was looking a lot more relaxed. "That's how I explained this to Gene. Does that help you, Paul?"

"Well, Norm, I can see now that opening air registers when a heater is caught up in a positive feedback loop is dangerous. The safe thing to do is get above the Point of Absolute Combustion first, by manually reducing the fuel gas flow, before increasing the combustion airflow rate."

"Exactly right," I concluded. "Air can be dangerous!"

Lessons Learned

- When a heater tube fails, use of the snuffing steam to avoid an explosion is mandatory.
- When a fired heater falls below its Point of Absolute Combustion, the heater's fuel gas flow must be immediately switched from automatic temperature control to manual, and then the fuel gas flow must be manually reduced until the heater outlet temperature starts to drop, as the fuel gas flow is throttled back.
- Do not switch back to auto if the heater outlet temperature goes up as you are cutting back on the flow of fuel gas.

11

Coking in Fired Heater Tubes: Safe Shutdown Due to Tube Failure

Introduction

In 1982, a visbreaker vacuum heater experienced a ruptured tube, requiring an emergency shutdown. The operators failed to use "snuffing steam" to preclude an explosion as the heater was taken out of service. In the resulting explosion, 40 operating and maintenance personnel were injured. The reason the snuffing steam was not used was that I had the snuffing steam lines blinded off because the snuffing steam was mainly water, and not steam.

Purpose

Combustion air has to be excluded from the firebox when the box's interior is above the explosive limit and is too fuel-rich to explode. As operators block in the fuel—and also the hydrocarbon feed to the ruptured tube—the firebox content drops from above the explosive region into the explosive region. This then results in an explosion, like the one that occurred in 1982 with the visbreaker vacuum heater.

Barry lay unconscious behind the partition. Burns covered his body where his polyester shirt had melted. Barry was my assistant and friend on the FCU-visbreaker complex at the Good Hope Refinery.

Heater Tube Failed

"Lieberman, get down to the visbreaker. We just blew a tube on the vacuum heater," Barry had screamed over the radio.

"What's happened?"

"We got black smoke and fire blowing out the heater stack and the sight ports. I got the guys putting fire water on the box. I told plant management before not to use fire water for cooling. In an emergency, we don't got enough water pressure!"

"Okay! I'll be right over. Don't forget the pass steam. We don't want to coke up the tubes."

Upper Explosive Limit

I stood on the Mississippi River levee, 200 feet back from the heater. Black smoke blew out of every sight port, and oily flames raced from the stack. Tom was blocking in the fuel gas and lining up the pass steam to the coils to blow out the residue before it coked up.

Barry was closing the 4-inch heater feed valves on all four passes and making sure the control valve bypasses were shut.

It looked bad. But, in reality, the heater was in a safe position, because the heater firebox was in a fuel-rich situation above its upper explosive limit. It was too fuel-rich to explode.

I knew this because I could see unburnt hydrocarbon vapors from the failed tube igniting in the flue gas that was racing out of the heater's stack. Of course, the concentration of hydrocarbons in the firebox would diminish with time as Barry blocked in the feed and closed off the fuel. And then, as the fuel-to-air ratio diminished, the heater box would drop into the explosive region. And then—the box would explode. I had seen this happen before.

Snuffing Steam

Barry should have introduced snuffing steam into the firebox at this time. This would have retarded the airflow into the firebox, and allowed the box to cool off in an air-deficient environment. That is, the steam would have raised the box pressure above atmospheric pressure and temporarily precluded airflow into the box due to draft.

But I knew Barry would not do this. A few months back, I had the snuffing steam lines blinded off from the visbreaker heater.

But why?

Because the snuffing steam lines were full of steam condensate. Their use would have blown water into a 1600°F brick box and created a pressure surge, which would have blown burning visbreaker tar all over the plant (Fig. 11–1).

I had suggested to Barry that, in the event of a tube failure, he instruct the operators to close off the air registers underneath the heater and shut the stack damper. This would prevent an explosive mixture of air and fuel from developing inside the firebox.

"Okay, Lieberman. In the event of a busted tube, I'll call you to close off the damper and air registers. I'm not gonna ask any of my guys to risk their lives."

Thus, the call from Barry!

As Barry completed closing off the visbreaker feed valves, a slight explosion shook the unit. He and 20 operators and maintenance personnel were scattered across the deck with their polyester shirts on fire. Only later did I learn about fire retardant Nomex coveralls.

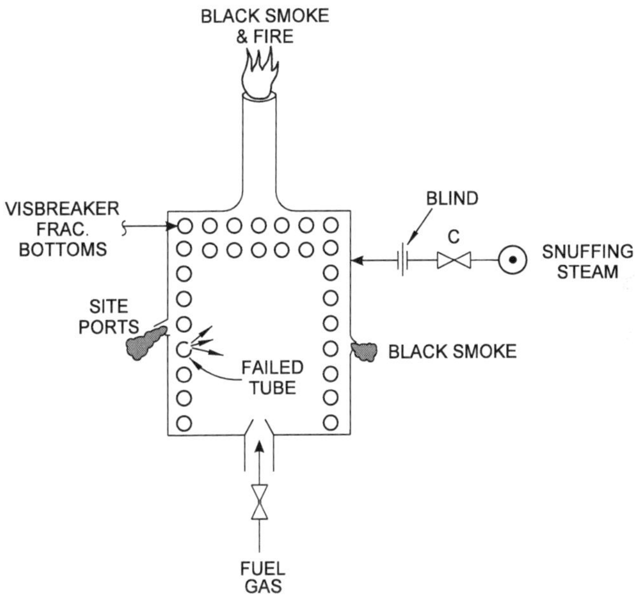

Fig. 11–1. Stuffing steam could not be used when tube failed. It had been blinded due to water in steam.

Abandoning Box Steam Installation

The above story happened just as I've explained. What is equally true is that the world's largest operator of refineries has disconnected snuffing steam from fired heaters for the same reason.

This came to my attention 10 years ago, in the course of instructing my "Process Equipment Troubleshooting Seminar" for this refiner at their Gulf Coast plant. The reason for this corporate-wide decision was the same as when I made my decision in 1980 at the Good Hope Refinery—that is, water contamination of the snuffing steam. The result has been periodic and repeated explosions in their fired heaters, due to tube failures, while shutting down the failed heater.

Corrective Action

Recently, another major operator of refineries shared with me their response to this chronic problem. I have shown their design for a snuffing steam installation in (Fig. 11–2). Here are the key features:

- Snuffing steam valve located at grade
- Steam valve located 100–200 feet back from the firebox.
- Lines steam traced and insulated
- Carbon rupture disks rated to rupture at 5 psig at the firebox snuffing steam inlet nozzles
- A half-inch drain hole at all low points
- "Car seal" open valves at the firebox inlet nozzles
- Upstream of the main snuffing steam supply valve, a condensate drain line, equipped with a steam trap
- Downstream of the main snuffing steam supply valve, a drain valve, car sealed open

Note that activation of this system only requires an operator to open one valve (valve "A") at grade. Leakage through this valve will be drained by the ½-inch weep holes (Fig. 11–2).

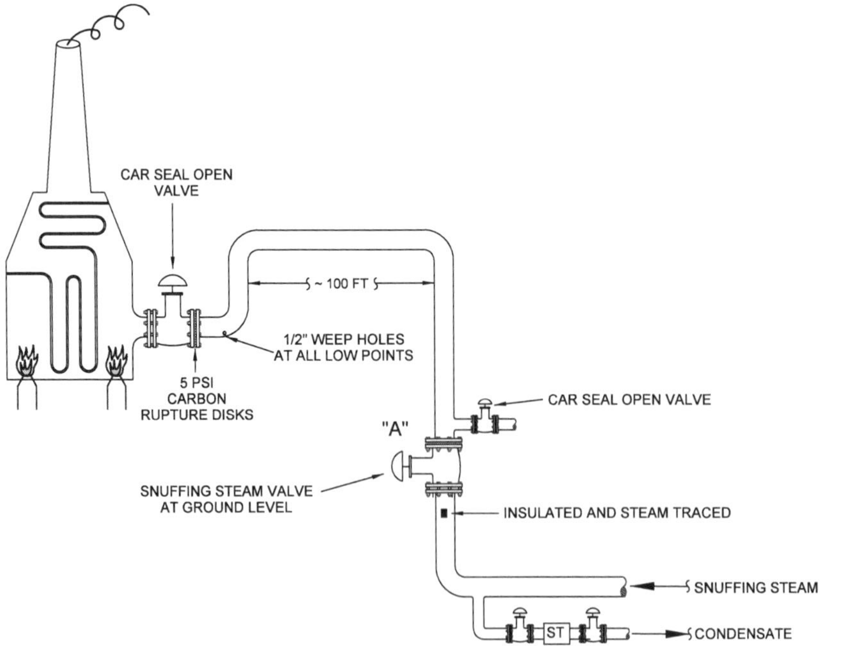

Fig. 11–2. Heater snuffing steam correct design

Typically, the snuffing steam valves I have seen are 6- or 8-inch carbon steel valves 20 feet up in the pipe rack, with chain operators. When I have tried to pull open such valves in an emergency, the valve was stuck due to its not being moved or greased in 20 years. All I accomplished was having the chain break off and hit me in the head.

A final note. Operators who are not trained will not know what to do in an emergency. Operator training consists of walking out to valve "A" (Fig. 11–2) and saying, "Look! When you got a fire in the heater, turn this valve clockwise."

Heater Tube Coking

The main problem that causes fired heaters in hydrocarbon service to shut down after just a few months of operation is coke accumulation inside heater tubes. The reason solid coke (that looks like coal) forms inside the tubes is *not* primarily due to such factors as these:

- Overheating tubes
- Flames impinging on the outside of the tubes
- Low linear velocity inside the tubes
- Tendency of the liquid hydrocarbons feed to decompose into coke

In my experience, the factor that causes solid coke to adhere to the I.D. of tubes is low tube side "mass velocity."

Mass Velocity

Linear velocity is expressed in feet per second. Mass velocity is expressed as the pounds per square foot per second. The "square foot" term is the cross-sectional area of the tube. Mass velocity is the same as momentum:

- Momentum = (Mass) • (Velocity)

The problem with a high-linear velocity, but a low-mass velocity, is that, at a low-mass velocity and high-linear velocity, the liquid and vapor phases can separate. The vapor rushes down the center of the heater tube. The liquid creeps slowly along the inside of the tube walls. Using more vapor (such as coil steam injection) makes the problem worse. The slowly flowing liquid in the tube's interior overheats, turns to coke, and plugs up the tube. This is what I have observed based on plant data accumulated over many years:

Table 11–1. Tube-side coking resistance in relation to tube mass velocity

Tendency to Resist Tube-Side Coking in Hydrocarbon Service	Tube Mass Velocity Pounds/ft²/second
Excellent	300–400
Good	200–250
Acceptable	100–150
Poor	60–80
Terrible	30–40

In one project, I revamped a refinery's delayed coker fired heater from four parallel passes to two parallel passes. This doubled the mass velocity from 50 to 100 $lbs/ft^2/second$. Heater run length increased from 3 months to a year. That was in 1971. But I recall the project (American Oil; Sugar Creek, Missouri) with pride and pleasure to this day. I design new Delayed Coker Heaters for a mass velocity of 400 $lb/ft^2/second$.

Lessons Learned

- Snuffing steam connections should always remain ready for service—never blinded.
- Wet snuffing steam creates its own hazards.
- Snuffing steam valves should be located at least 100 feet back from the fired heater.
- Snuffing steam connections should be isolated from the heater firebox, with rupture disks designed to fail with a few psi of steam pressure.
- Snuffing steam isolation valves should be located at grade and easily accessible during an emergency.
- Grease valve stems once every year.

12

Centrifugal Pumps

Introduction

A centrifugal pump, at normal flows, pumping a low-viscosity fluid, driven by a constant speed motor, develops the same feet of head, regardless of the specific gravity of the liquid being pumped. The pump impeller does not increase the fluid pressure; it just increases the fluid velocity. The pressure inside the pump case is the pump's suction pressure and not the pump's discharge pressure. This chapter only deals with fixed-speed centrifugal pumps—not variable-speed drives or positive-displacement pumps.

Purpose

The most important concept in this chapter is an understanding of the factors affecting the integrity of the pump's mechanical seal. "Packed" pumps using a "stuffing box" have been rarely used in the industry since the 1980s, except for pumping water. Adjusting the pump's seal pressure too high will damage the seal, which is very costly to repair or replace. The correct seal flush pressure for centrifugal pumps with an external source of seal oil is slightly (10–20 psi) above the pump's suction (not the discharge) pressure.

"Louis, using 100-psig seal flush pressure is excessive," I explained.

"The pump discharge pressure is 80 psi. The seal flush pressure should be above the discharge pressure," Louis, the operator, insisted.

"No," I said, "Seal flush pressure must be only a few psi higher than the suction pressure. As this pump is in reduced crude service, and the flush to the seal is diesel, we're losing a lot of diesel to reduced crude. For older pumps, you can calculate the amount of seal fluid lost into the product as follows:

- 3 GPM per inch of shaft diameter, when the pump is running
- 50% of that amount when the pump is idle
- 200% of that amount if there is both an inboard and outboard seal
- For more modern pumps, with cartridge seals, use one third of above

- If seal flush pressure exceeds design pressure, because it's running above the pump's discharge pressure, the amount of seal flush consumed may get really high in proportion to the following:

  Seal Oil Flow $\approx (P_{seal} - P_{suction})^{½}$

where:

  - P_{seal} = Seal Oil Pressure downstream of the seal flow adjustment valve, psi
  - $P_{suction}$ = Pump Suction Pressure downstream of the suction strainer, psi."

"What you say makes sense, if the pressure inside the entire pump case is the suction pressure, rather than the pump discharge pressure. But the pressure of the liquid inside the pump case is certainly the discharge pressure, because the liquid has been pressured-up by the pump impeller (Fig. 12–1)," Louis explained.

"What you're thinking does apply for a positive displacement pump. However, most of the pumps we use are centrifugal pumps, where the function of the impeller is not to increase the process liquid pressure, but to accelerate the flow," I responded.

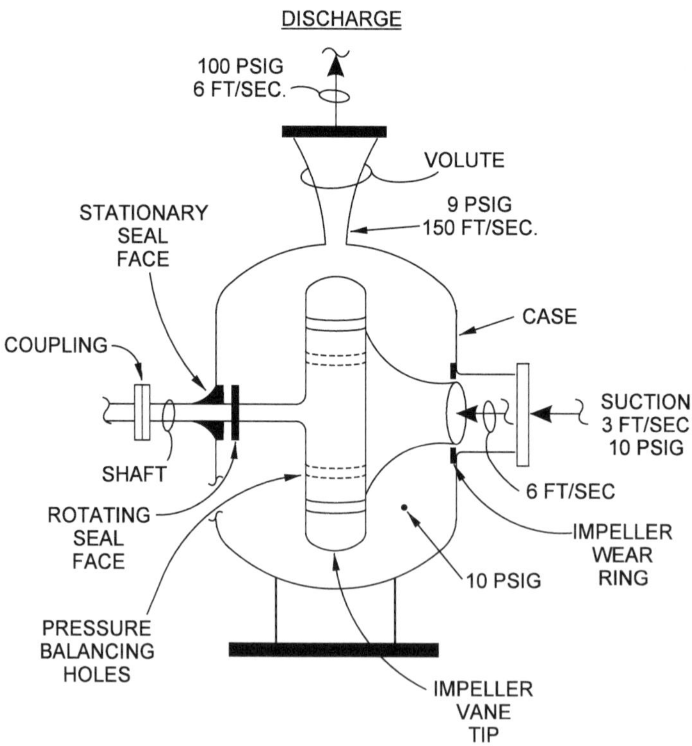

Fig. 12–1. Pressure inside the pump case is the pump suction pressure. Pressure increases as liquid flows thru volute.

Velocity, Head, and Pressure

I once performed an experiment as a teenager. I had been told that the water pressure in our bathroom was 40 psig, and I wanted to measure it. So, I took our bathroom scale into the backyard. Next, I adjusted the hose's nozzle to get the biggest pressure possible from the hose. Then, I directed the jet of water straight down onto the scale.

Much to my disappointment, the scale reading did not increase to 40 pounds, but only to about 6 pounds. But then I noticed that the jet of water was concentrated in a circle of only a half-inch diameter.

One-half inch is equal to an area of:

- $(0.5)^2 \times \pi/4 = 0.2$ square inches

So I then divided the 6 pounds reading on the scale by 0.2 $(\text{inch})^2$:

- $6 \text{ lbs} \div 0.2 (\text{inch})^2 = 30 \text{ lbs}/(\text{inch}) = 30 \text{ psi}$

I figured that the missing 10 psi (40 – 30 = 10) must be due to losses, because the water had to flow a mile from the water tower on Veteran's Blvd. to my house.

I could understand that loss of 10 psi. But what I found confusing was that the water flowing from the end of the garden hose nozzle was not at a pressure of 40 or 30, but at 0 psig, or atmospheric pressure. So I wondered—from where was the 30-psi pressure that I had measured on my bathroom scale coming?

Conversion of Velocity to Flowing Pressure

According to Daniel Bernoulli, there are several forms of energy:

- Heat—Thermal energy
- Velocity—Kinetic energy
- Head—Potential energy
- Pressure—Effect of gravity
- Electricity

Directing a jet of water from a garden hose straight-up converts the high velocity (kinetic energy) of the water into feet of head (potential energy). At sea level in New Orleans,

- $DH = 0.178 \bullet V^2$

where:

- DH = Height of water, inches
- V = Velocity of water, ft/sec.
- The 0.178 is the theoretical conversion factor for converting the kinetic energy or velocity of water (ft/sec) to inches of water head (i.e., potential energy).

If the water, shooting out of the end of a garden hose at 30 ft/sec, was directed straight-up, it would rise to:

- $0.178\,(30)^2 = 160$ inches ≡ 13½ feet

Actually, the water would not quite rise to 13½ feet, because of frictional losses.

Function of a Centrifugal Pump Impeller

"Let's refer to Fig. 12–1 again. A typical suction piping velocity might be 3 feet per second. Then the liquid flows into the eye of the pump's impeller and accelerates to, say, 6 feet per second. That acceleration of the flow from 3 to 6 feet per second requires energy."

"Does that acceleration energy come from the pump's impeller?" asked Louis.

"No. It comes from the potential energy of the liquid, or from the head of liquid."

"Norm, is that what you call the *'Required NPSH' of the pump*?"

"Yes."

"And if the required NPSH isn't smaller than the head of liquid available in the vessel I'm pumping out of, then I need to raise the vessel liquid level. Otherwise, my pump's going to cavitate," suggested Louis.

"Yes. You will damage the pump's mechanical seal due to vibration."

"What happens to the liquid after it flows into the eye of the impeller?"

"The liquid is spun around inside the impeller. As shown in Fig. 12–1, the liquid accelerates from 6 feet per second to 100 feet per second as it discharges from the vane tip of the impeller."

"Norm, where does the energy to accelerate the liquid come from? Also from the available NPSH?" asked the operator.

"No. It comes from the electric motor, or steam turbine, spinning the impeller. Electricity, which is a form of energy, can be converted to kinetic energy, which is what's happening inside of the pump's spinning impeller.

"The pressure of the liquid discharged from the impeller is a little bit lower than the pump suction pressure. The impeller discharge pressure is the pressure inside the entire pump case. Just to make really sure that the pump's seal doesn't see any pressure much above the pump suction pressure, there are holes (pressure-balancing holes) usually drilled through the back of the impeller, as I've shown in Fig. 12–1"

"I've seen those holes before," said Louis. "Carl told me those holes were there to keep the pressure on both sides of the impeller equal."

"Carl's right. But that pressure he's referring to is the pump suction pressure, not the discharge."

"None of this makes any sense. How does the pump discharge pressure increase from the 10-psig suction to 100-psig discharge? I'm still confused," Louis wondered.

Converting Velocity to Head

"Let's look down into the discharge nozzle of an unflanged pump," I suggested. "The discharge flange is 6 inches, but the opening down inside the pump case is a lot smaller."

"It looks like it's a 1½-inch opening."

"It follows, then (Fig. 12–1), that the velocity at the 6-inch discharge flange has dropped from 100 feet per second down to 6 feet per second:

- 100 ft/sec • $(1.5/6)^2$ = 6 ft/sec

"And the reduction in velocity has been converted into—"

"Hey! I got it!" Louis exclaimed. "It's just like a roller coaster car slowing down as it goes uphill. The speed of the car is converted to elevation, or what you call feet of head. Or what Bernoulli calls potential energy. But that doesn't happen until the liquid escapes out of the pump itself. That's why the seal flush pressure only needs to be a little bit higher than the pump's suction pressure. The discharge pressure really doesn't affect the pressure that the pump's mechanical seal has to hold. And even if the seal flush pressure did build up a bit behind the impeller, that pressure would bleed itself back to the pump suction through pressure-balancing holes drilled in the back of the impeller."

Relationship of Head to Discharge Pressure

"Norm, you talk about 'head.' As an operator, I only see pressure—not head. Doesn't the centrifugal pump create a certain amount of discharge pressure?"

"A centrifugal pump, which is pumping a nonviscous product, develops a certain amount of feet of head at a given volume of flow, regardless of the density or specific gravity of the liquid pumped. To convert from feet of head to pressure—which is what you're interested in—we use the equation:

- $DP = \dfrac{(DH) \bullet (SG)}{2.31}$

where:

- DP = Increase in discharge pressure over suction pressure, psi.
- DH = Increase in feet of head developed by pump, feet.
- SG = Density of liquid at flowing temperature compared to density of cold water at 60°F. (That is, Specific Gravity).

"So a pump that's pumping diesel will put up more discharge pressure than a pump that's pumping naphtha, because the diesel's denser than the naphtha?" Louis asked.

"Correct."

"I got the specific gravity idea. But how about the 'nonviscous' idea?"

"Most of the liquids we handle are low viscosity (less than 50 centistokes). The only high-viscous material is cold heavy crude, or cool vacuum tower slop wax. Anything that has a viscosity of more than 100 centistokes wouldn't be handled with a centrifugal pump, but a different type. That is, a positive displacement pump."

"Let's just stick with the centrifugal pump for now. You can explain about positive displacement pumps later. But how about what you just said about 'At a given volumetric flow'? What does that mean?"

Pump Curve

"Let's refer to Fig. 12–2. The horizontal axis of the charge is flow in GPM. The vertical axis on the left side of the chart is head in feet. The vertical axis on the right side is required horsepower. It doesn't matter if you're pumping water with a specific gravity (S.G.) of 1.00 or of gasoline with an S.G. of 0.75. The pump will still produce the same feet of head, as long as the GPM, or flow, is constant."

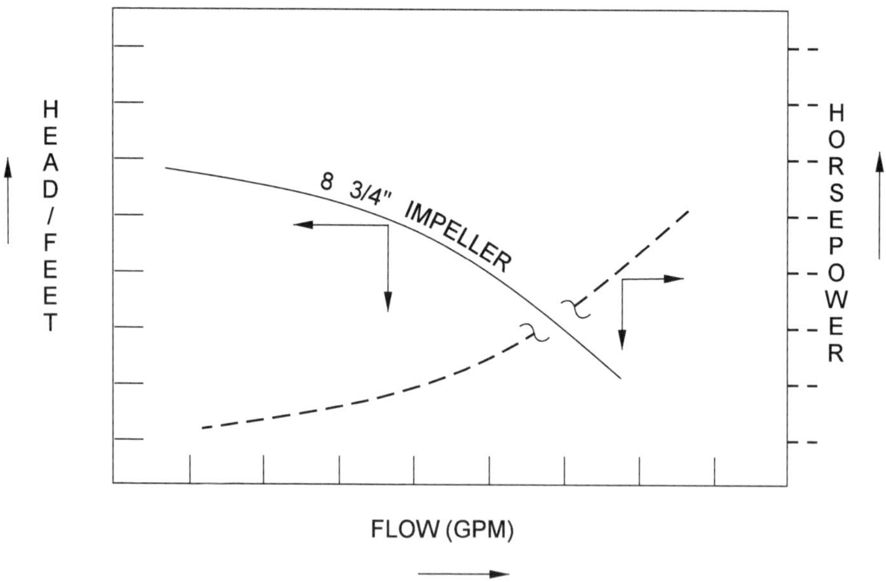

Fig. 12–2. Pump performance curve. Amps are proportional to motor horsepower.

"But how about that horsepower curve? Is that the same for water and gasoline also?"

"No, Louis. The motor horsepower, or the amount of amps the motor will pull, is a linear function of the S.G. If the S.G. increases by 20%, so will the motor amps. The pump curves are developed using water in a test rig. Even though a pump may be designed for diesel with an S.G. of 0.85, the horsepower shown on the right-hand vertical axis of the pump curve is likely for water."

"That explains why the P-109 pump trips off on high motor amps, when I switch from water to caustic, unless I throttle back on the discharge of that pump first."

"Yes. The S.G. of the water is 1.00. The S.G. of the 20 Baume caustic is 1.20. At the larger S.G., you're going to be limited by motor horsepower. Not the pump's capacity."

Positive Displacement Pumps

"Norm, I have one more question. Out in the tank field, we have a crude charge pump that is different from our centrifugal pumps. I looked at it a while ago; it looks the same as our centrifugal pumps. What's the difference?"

"Both types of pumps do appear the same. But that pump in the tank field is likely a positive displacement pump. The difference is this:

- Centrifugal pumps develop the same feet of head for a particular flow rate, as shown on their pump curve.
- Positive displacement pumps do not run on a curve. They put up as much head or discharge pressure as they need to overcome the pressure of the system they're pumping into. If the pump needs to push 100,000 BSD of crude into a nearby tank, its discharge pressure might be just 25 psig. If the pump needs to push 100,000 BSD of crude to your distant crude unit, its discharge pressure could be 150 psig."

Seal Oil Flow to Idle Pumps

"Thanks. I'm going to re-set the seal flush pressure on the pumps, just like you told me. A bit above the pump's suction pressure. Does that go for the spares also, that I'm not operating?"

"No, Louis! For example, take the resid pumps that use diesel for seal flush. When you shut the pump down, wait a few minutes for the diesel oil to flush through the pump. Then block the diesel seal flush in. Even though the pump is not running, the seal flush flow will still be around 60% of its flow when the pump was in operation. Every barrel of diesel you lose into resid is costing the refinery 5 to 10 dollars. So let's not waste seal flush on pumps that are not in operation."

"I've been leaving the seal flush lined up to the idle pumps for years. Someone should've told us not to do that. What a waste!"

Safety Note

The heavier the material, the lower its auto-ignition temperature, hydrogen being an exception. Gasoline has an auto-ignition temperature of 480°F. Asphalt, when spread over a porous surface, like your clothes, auto-ignites at 300°F.

That's why a seal leak on a heavy vacuum gas oil pump is so dangerous. It's much safer to get a sample of naphtha at 350°F, than vacuum resid at the same temperature.

Benefits of Excess Available NPSH

Supplying extra "available net positive suction head" tends to suppress formation of bubbles, and hence of cavitation, inside a pump's impeller. This will improve the reliability and safety of the pump, by extending the operational life of the pump's mechanical seal. Also, seal maintenance costs will be reduced and will offset the capital investment for the increased vessel elevation. A new double-mechanical seal may cost sixty thousand dollars to purchase.

Lessons Learned

An ordinary pump, driven by an alternating, three-phase motor, runs at a constant speed, regardless of the load on the pump. An ordinary pump discharge pressure increases with the S.G. of the liquid pump, even though the feet of head developed by the pump is constant, provided the flow rate is constant.

When increasing the impeller size, care must be taken to avoid exceeding the Maximum Allowable Discharge Pressure (M.A.W.P.) of heat exchangers not protected by relief valves.

13

Starting NPSH for Centrifugal Pumps

Introduction

Centrifugal pumps tend to cavitate on startup because the fluid in the suction line must be accelerated. The energy to accelerate this liquid does not come from the pump. It must come from the available "net positive suction head." I call this the starting NPSH of the pump.

If the suction line is very long, this may be a big problem in that the pump will cavitate on startup, unless the operator opens the pump's discharge valve very, very slowly.

Purpose

The available NPSH used for design must take into account these factors:

- Frictional loss in the suction line
- Acceleration in the suction line
- Nozzle losses in the suction line
- Changes in flow in the suction line

Joe Johnston was the operations manager for a refinery in Casper, Wyoming.

"Norm, I can't get the new flashed crude charge pump started. It cavitates as soon as I push the motor start button."

"Have you tried raising the liquid level in the crude flash drum to maximum?"

"Yes. That did help. But not enough to get the pump started without cavitating. I don't want to start up and ruin the $70,000 double-mechanical seal."

"Have you opened the spill-back on the 3-inch discharge to suction line?"

"I tried that. The pump suction got warmer, and it made my problem worse. After all, required NPSH increases as flow increases." (Fig. 13–1)

"Let's spray cold water on the suction line and pump case, Joe. That will subcool the flashed crude by a few degrees and generate more available NPSH by reduction of the vapor pressure of the flashed crude."

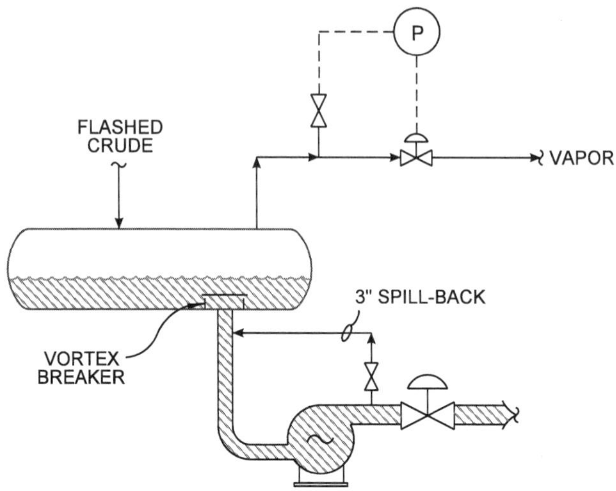

Fig. 13–1. Typical pump installation

"Absolutely not! I'm not going to tear off $20,000 of new insulation. I need to get that pump running today. I know that the flashed crude drum is not elevated high enough. It's a design error. Needs to be like another 10 or 20 feet higher above the pump suction to provide more NPSH for startup."

"Let's try opening the discharge valve a fraction of a turn, start the motor, and then open the discharge valve real slowly," I said.

"I've tried that. It helps some. But still the pump cavitates. I need a lot more NPSH to get that pump up and running sometime today."

"Okay, Joe. Let's go to lunch. I'll explain how to get the pump started without cavitation, and I'll also tell you a story about my daughter who studied engineering."

The Nature of NPSH

"Dad, we learned about pump NPSH today. Professor Albright explained it. It's real simple."

"Actually, Irene, it's not that simple. It's a complex subject."

"No, Dad! It's really, very simple. I'll explain. First, you've got your required NPSH. You read that off the manufacturer's pump curve (Fig. 13–2). As the flow goes up, the required NPSH always increases.

"Then you've got your available NPSH. I learned how to calculate this yesterday:

- Available NPSH = $[(P_1 + P_A) - (VP)] \cdot 2.31 \div S.G.$

where:

- NPSH = Feet
- P_1 = Pressure at pump suction (psig)

- P_A = Atm. pressure, psi
- VP = Vapor pressure of liquid at pump suction, psi
- S.G. = Flowing specific gravity

"All you do is provide enough available NPSH so that it's equal to or exceeds the required NPSH at the max design flow rate of the pump. Professor Albright says that you can get enough of that NPSH by either elevating the drum that the pump is serving or subcooling the liquid that is being pumped."

"Subcooling, Irene?"

"You know, Dad. Maybe like spraying icy cold water on the pump's suction line. That'll reduce the liquid vapor pressure below its saturation temperature."

Nozzle and Piping Losses

"There are three factors, Irene, that increase required NPSH beyond that which you learned at the university and beyond that specified on the manufacturer's performance curve.

"One of these factors is nozzle losses. Another is piping friction. For low viscosities—less than 20–30 centistokes—the nozzle exit loss is:

- $DH_1 = (0.3) \bullet (V)^2 \div 12$ (2)

where:

- DH_1 = Loss of NPSH due to acceleration through the nozzle, feet
- V = Nozzle exit velocity, in feet per second."

"I made a note, Dad. Good point."

"The 0.3 coefficient assumes, Irene, that there is no vortex breaker on top of the nozzles (Fig. 13–1)."

"But, Dad, I think most nozzles in refinery process vessels have vortex breakers. What coefficient should I use with a vortex breaker?"

"As a guess, you might increase the 0.3 to 0.5 coefficient in equation (2). But, I'm not sure. "And then there's suction piping losses due to friction. You can calculate the piping losses using this equation:

- $DH_2 = \dfrac{4}{\text{I.D.}} \bullet \dfrac{DL}{62.3} \bullet \dfrac{(V)^2}{27.7}$

where:

- DH_2 = Piping loss per hundred equivalent feet, psi
- DL = Density of fluid, lb/ft^3
- V = Line velocity, ft/sec
- I.D. = Pipe diameter, inches

You can convert pressure loss from psi to feet by using this formula:

- $(DH_2) \bullet (2.31) \div$ Specific Gravity

I've assumed an ordinary roughness factor inside the pipe in this calculation."

"But Dad, how do I know the piping equivalent feet, before the plant piping is designed?"

"As a guess, I would use the actual distances between the pump and vessel, and double it."

"But you've guessed also at the roughness factor inside the pipe. Dad, I think that roughness is used to calculate the friction factor. If you've guessed at the friction factor, you've guessed at the suction piping losses. That doesn't seem professional! Engineers aren't supposed to guess."

The Bhopal Factor

"You're right, Irene. Engineering is a serious business. People die when we're not careful. You've heard about the disaster in Bhopal, India, in 1984?"

"No. That was before I was born. Is this going to be one of your long stories?"

"Over 10,000 people were killed. There was someone from Bhopal in a seminar I instructed recently."

"Was the cause of those people being killed...was that because of a pump not working...like it cavitated? Like it didn't have enough NPSH?"

"Well, yes, but only indirectly. First, I'll have to tell you a bit more about cavitation. It's not simple."

Symptoms of Cavitation

"Irene, what have you been taught are signs of cavitation in a centrifugal pump?"

"Dad, that's simple. The pump discharge pressure is erratic and low. Pump flow is also erratically low. Mainly, the pump will make a rattling sound, like shaking a bunch of bolts in a can. Also, operating a pump in a cavitating mode will damage it. The impeller or case will be damaged, maybe? But, as long as the required NPSH is a little bit less than the available NPSH, cavitation will be totally avoided."

"Irene, most pumps that I find cavitating do not have a highly erratic discharge pressure, unless the suction line or draw-off nozzle is partly restricted. Certainly, discharge pressure may be only 50–70% of the pump's performance curve head. But the discharge pressure is often only slightly shaky, and the pump runs quietly.

"I call this 'marginal cavitation,' in the sense that if I raise the pump suction head by just a few feet (i.e., 1 or 2 psi), the discharge pressure may increase from 120 psig to 180 psig. In marginal cavitation, a pump appears to be running on an inferior performance curve; a bit more available NPSH restores it to its proper performance.

"The problem is that, even when a pump's calculated available NPSH exceeds the curve-required NPSH by a few feet, fluid as it flows through the pump's impeller may still partly vaporize. The bubbles of vapor can damage the pump's mechanical seal's soft carbon face. This, with time, will cause the seal to leak."

"I see. But how can that be prevented? That sounds serious."

"Well, Irene, you have to design the pump suction for extra NPSH, beyond that minimum specified by the pump performance curve."

"But how much extra NPSH is needed to make sure the seal isn't damaged due to cavitation?"

"My pal Ken Bloch wrote a book on this subject.[II] Want to read it?"

"No! Is that what happened in Bhopal? The pump seal blew out?"

"Not exactly. The pump seal on their pesticide plant leaked repeatedly. So, the pump was bypassed, and they used nitrogen pressure to push the toxic reactant into the pesticide reactor. But, in bypassing the pump, they also bypassed their feed tank high-temperature alarm. So, when they accidentally got water in the feed tank, it reacted with one of the pesticide chemicals and formed a deadly gas. But, without the high-temperature alarm, the plant operators didn't realize there was a problem. If the pump had been designed with extra available NPSH, this wouldn't have happened."

"Maybe I will read your friend's book. But, Dad, is that the only reason for extra available NPSH when designing a new centrifugal pump?"

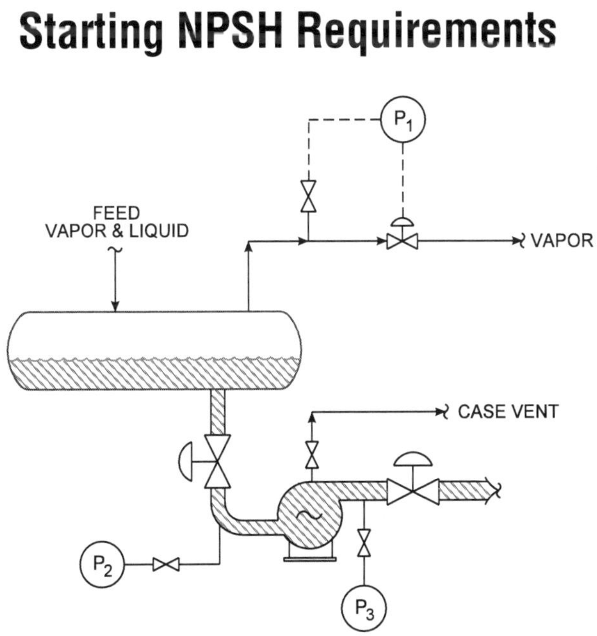

Fig. 13–2. A sudden increase in drum pressure temporarily increases available NPSH for start-up.

II. Bloch, Kenneth, *Rethinking Bhopal*, Elsevier, 2016.

"No, Irene. There is a more fundamental reason for supplying available NPSH in excess of the manufacturer's NPSH, which is tabulated on the pump performance curve. Let's take a look at (Fig. 13–2)."

"I'm ready, Dad. You talk and I'll study (Fig. 13–3)."

"Well, it's complicated. I'll explain step-wise. Initially, when an operator starts a centrifugal, motor-driven pump, he 'primes' the pump by taking these steps:

1. Fully opening the suction valve
2. Partly opening the vent on the top of the pump case
3. Opening the discharge valve just a little bit. Meaning, turning open (that is counter-clockwise) the discharge valve about one sixth of a full revolution (i.e., one spoke)
4. Pushing the motor start button
5. Then, opening the discharge valve the rest of the way, as fast as possible, but as limited by not losing the pump ***suction*** pressure (P_2 in Fig. 13–2)."

"But, Dad, wouldn't it be better to open the valve slowly. Why rush?"

"No, Irene. Running a centrifugal pump at a very low rate causes excessive vibrations, which is bad for the pump's bearings and seal."

"But, Dad, I learned at university that it's also bad for the pump if it cavitates. So it seems to me that it's best to watch the discharge pressure. And, if the discharge at P_3 (Fig. 13–3) becomes erratically low, open up the discharge valve more slowly."

"You're right. If the pump cavitates, it will damage the pump's bearing, but especially the mechanical seal. Most of the pumps used in process plants are 'self-flushed pumps.' That means the seal flush fluid comes from the discharge of the pump itself. If the pump cavitates, the seal flush flow stops. Then, the seal loses lubrication and cooling. Better to watch the suction pressure. By the time the discharge pressure slips down, the pump is already cavitating, and seal damage is occurring. By watching the suction pressure, you can avoid cavitation and seal damage before the problem starts."

"Very interesting. Does that mean, Dad, that there is always a pressure gauge located on pump suctions?"

"Unfortunately, no, Irene."

"Then how can the plant operators watch the suction pressure?"

"Here's how. Process pumps almost always have a spare pump piped in parallel with the main pump. What I do is watch the discharge pressure of the spare pump that is not in operation (with its suction valve open and discharge valve shut). This will be the same as the suction pressure of the pump that is running."

"Why is it that pumps cavitate so much when they're being started, but not so much when they are running?"

"Pumps cavitate while being started because of their '**starting NPSH required**.'"

"Is that the same as required NPSH? Or is that something different?"

"Look at (Fig. 13–2), Irene. When a pump starts, stagnant liquid in the suction line has to be accelerated to maybe 3 feet per second. Here's the point:

"Centrifugal Pumps Push; They Do Not Suck."

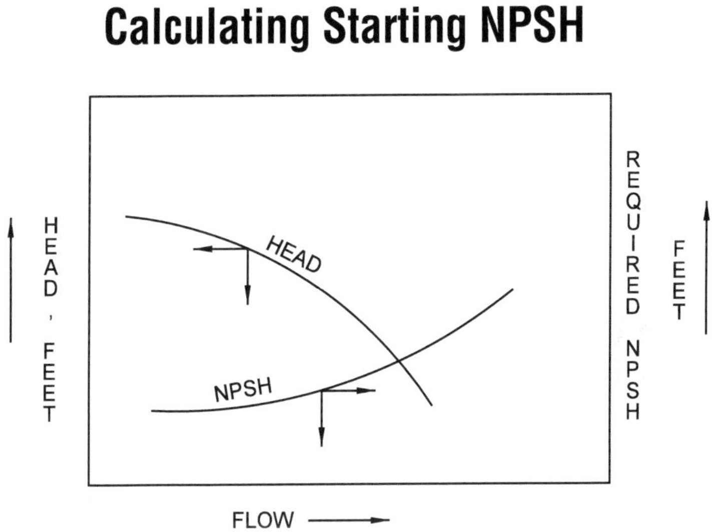

Fig. 13–3. Manufacturer's pump curve showing required NPSH.

Anyone who has commissioned a fixed-speed, motor-driven centrifugal pump will have noted the need to increase the flow slowly. This is not a problem on a variable-speed turbine drive. Then, the turbine speed is increased slowly. The problem is the need to supply energy to accelerate the fluid in the line feeding the pump. If the suction line is short, this is a minor problem. For example, if I am going to switch to the spare pump, 90% of the suction piping is already flowing. I can then start the idle spare pump with the discharge valve wide open.

However, if the suction line is long, then I have to provide sufficient energy to accelerate the mass of liquid in the suction line. This energy does not come from the pump. It comes from the static head of liquid in the suction line. That is, it comes from the available NPSH.

To summarize, centrifugal pumps require three sorts of NPSH:

- The NPSH, as shown on the pump curve Fig. 13–3
- Hydraulic losses in suction piping and nozzle exit losses
- Acceleration of liquid in the suction line

I've called this last item, which is often larger than the first two, the 'starting NPSH requirement.' It's a function of both the length of the suction line and the speed at which the plant operator opens the pump discharge valve.

The need to accelerate fluid in suction of a pump is called the starting NPSH requirement. To calculate starting NPSH requirement, assume the following (Fig. 13–4, 13–5):

- Suction line = 100 ft.
- Line is 8 inches.
- Fluid = water @ 62 lbs/ft.^{3}
- Initial velocity is zero.
- Final velocity is 10 ft/sec.

Then proceed as follows:

- Mass in line is 2,170 pounds.
- Change in momentum in line is 10 ft/sec times $2{,}170\text{lbs} = 22{,}170\dfrac{\text{ft.lbs}}{\text{sec}}$ (m)

If the operator increases flow uniformly over 100 seconds, the change in momentum is:

$$22{,}170\frac{\text{ft.lbs}}{\text{sec}} + 100\,\text{sec} = 222\frac{\text{ft.lbs (m)}}{\text{sec}^2}$$

Then, one pound mass equals one pound force times sec^2/ft. Therefore, force needed to accelerate water in pump's suction is:

- 222 lbs (f).

The area of the 8-inch suction line is 50 square inches. Therefore, starting NPSH required is:

- 222 lbs (f) ÷ 50 inch^2 = 4.4 psi

or:

- (4.4 psi) • (2.31) = 10 feet of water.

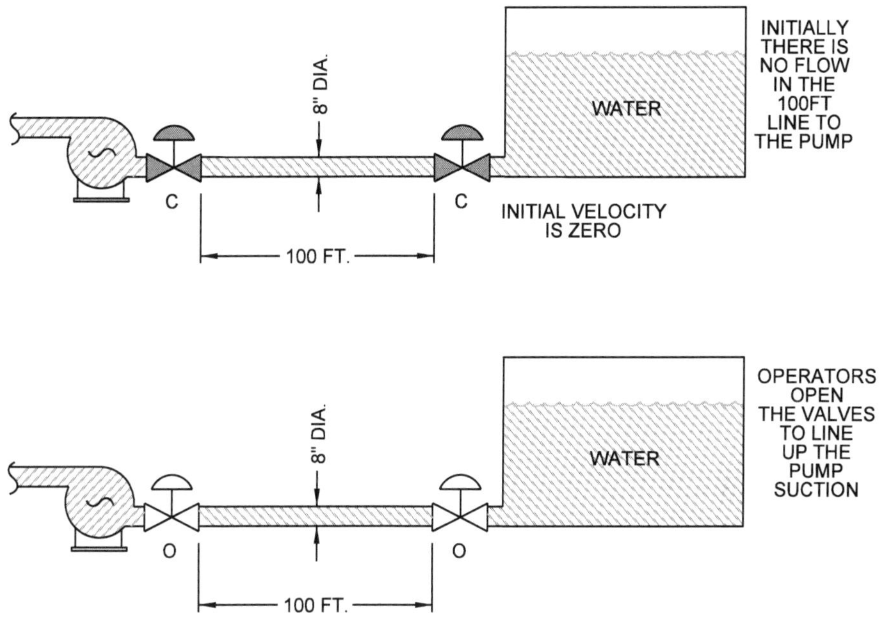

Fig. 13–4. Pumping water from a tank

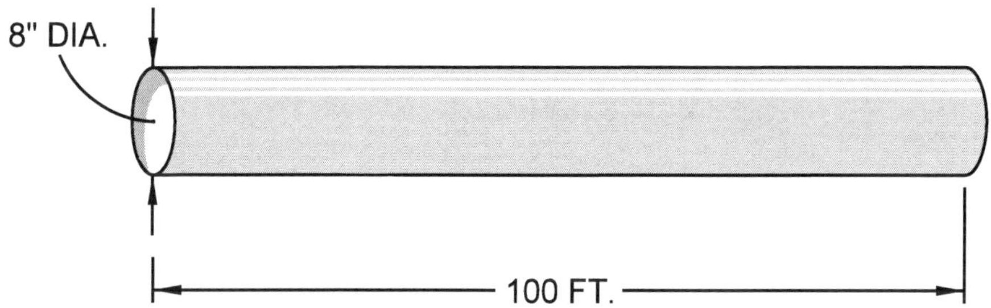

Fig. 13–5. Volume of 100ft, 8" diameter feed line

Increasing Process Flow Rate

"Dad, suppose a pump is running close to its design available NPSH. Then, the operator decides to increase the flow by 20%. The required NPSH will increase a lot because the velocity in the suction line will also increase, or accelerate, by 20%. Then the pump will cavitate, the seal flush will stop, and the seal will be damaged due to lack of lubrication. That's another good reason to design pumps with a lot more available NPSH than the manufacturer's pump curve shows."

"Yes, Irene. The engineer needs to take these factors into account when deciding on the elevation of vessels. Failing pump mechanical seals are expensive—also dangerous."

Increasing Available Starting NPSH in Casper, Wyoming Refinery

"Joe, first we'll increase the drum set point pressure by 10 psi.

"Next, we'll open the discharge valve (see Fig. 13–2) by one quarter of a turn.

"Then start the pump, and open the discharge valve, in just a few minutes, all the way," I said.

"No, Norm. That can't help. The suction pressure will increase by 10 psi. But the vapor pressure of the flashed crude will increase by the same 10 psi. According to your equation (1), the increased vapor pressure of the flashed crude will offset the higher suction pressure."

"Joe, after 10 minutes, what you say is true. But, at first, due to the 10-minute residence time of liquid in the drum, the higher vapor pressure liquid, formed at the vapor-liquid interface, will not migrate into the pump suction for that 10 minutes. If you can get the pump up and running quickly, you will have a temporary surge of available NPSH.

"Joe, it's a startup trick that I've used. Best to design the equipment elevation properly and allow for the starting, as well as the running, NPSH requirements. Often, while operators are learning this trick, they repeatedly cavitate the pump and damage the $75,000 seal."

Practical Steps to Minimize Cavitation

"Joe, these are steps to protect mechanical seals from damage due to cavitation:

1. Place a pressure gauge with a red mark on the pump suction. Train the operators to open the pump discharge valve as fast as possible, consistent with the pressure not falling below this red mark.
2. Raise the suction drum pressure rapidly by a few psi, just before starting a pump. Then reduce the pressure very slowly once the pump is operating.
3. During normal operations, increase flows really slowly. Most experienced operators tend to do this by themselves.
4. Run near the maximum safe drum level.
5. Spray cold water on the pump case and suction line to subcool the liquid.
6. Running two pumps in parallel requires a lot less NPSH—but is often difficult to control.
7. Injecting a subcooled liquid slip-stream into the pump suction will increase available NPSH.
8. A discharge spill-back to the suction line diminishes the available NPSH and also increases the required NPSH.

9. For steam turbine-driven pumps, increase turbine speed slowly.
10. Maintain pump suction screens and keep them clean. The openings in the screen are set by the pump's mechanical seal clearance for self-flushed pumps.

"Joe, keep in mind that pump seal failure is a major cause of fires and explosions in process plants. Especially when pumping propane and butane. Make sure the high-pressure and high-level alarms on the outer seal flush reservoir pot (for pumps with double-mechanical seals) are working, to alert operators when the inner seal has failed."

Safety Note

The double-mechanical seal is a safety device. It's not for reduction in environmental emissions. If the high-pressure or high-level alarms on the reservoir pot are not working, the safety benefits for the double-mechanical seal are largely lost. Always remember that an alarm or trip that is not tested routinely will never work in an emergency.

Lessons Learned

The elevation of vessels feeding centrifugal pumps should include an allowance for the need to accelerate the fluid in the suction line, not only for startup, but also for rapid changes in flows during normal operations.

The incident where 10,000 people were killed in Bhopal, India, when Union Carbide failed to take this factor into account, shows the importance of supplying an extra amount of NPSH to preserve the integrity of the pump's mechanical seal. This incident led directly to the demise of Union Carbide, the largest chemical company in America in the late 1980s.

14

How Controls Work

Introduction

The following parameters are controlled in a process plant:

- Temperature
- Pressure
- Level
- Flow
- Viscosity
- pH
- Boiling range
- Color
- Flash point
- Density
- Speed

Purpose

The first four parameters listed in the introduction are the most critical. The most difficult to measure, for control purposes, are levels. Level indication is influenced by plugged level taps, moisture in level-sensing devices, changes in specific gravity, and vaporization in the level-sensing chamber.

Measuring flows can also be difficult due to changes in the fluid specific gravity and plugged orifice taps. Control valves are subject to a wide range of malfunctions. Only too often, a control valve is left on manual control—but it should be in its automatic mode.

I have five senses that keep me in contact with my surroundings: sight, hearing, smell, touch, and taste. Most process plants have only four senses: temperature, pressure, level, and flow. Understanding how these parameters are measured, and their malfunctions, is an important part of working in a chemical, fertilizer, or refinery facility. It's rather like cooking; without a sense of smell or taste, our culinary efforts would not succeed.

Temperature

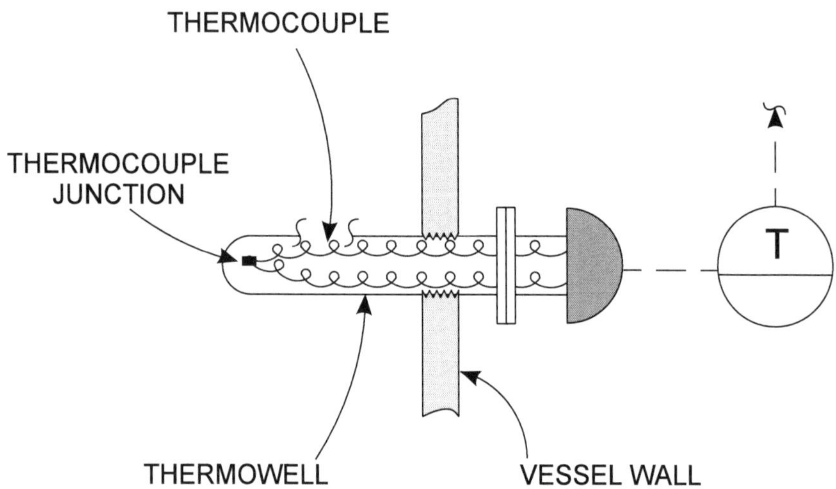

Fig. 14–1. Thermowell assembly

Temperatures that are transmitted to a remote location, for control purposes, are usually measured in this way:

- Take two wires of different metallurgy.
- The bare ends of each wire are twisted together to form a ***junction***.
- When this ***junction*** is heated, a tiny, direct-current electrical flow is generated.
- The hotter the junction, the bigger the current flow.
- For a given set of metal types of the wires, a unique flow of current is generated for a temperature. The temperature readings that are transmitted are really just tiny flows of electric direct current that increase with temperature.

The two very thin wires, with only their bare ends twisted together, form what is called a thermocouple. To protect the thermocouple from damage, it's inserted into a 6-inch-long, ¾-inch stainless-steel pipe called a thermowell (Fig. 14–1), which is screwed into the process pipe or vessel.

The thermocouple can be safely extracted from the thermowell, without danger. However, unscrewing the thermowell from a vessel while it is in service can result in a fatality. I have myself made this error on several occasions.

The other common method that is used to measure temperatures is by an optical pyrometer. That is, light radiation emitted by surfaces varies with surface temperature. However, in the range of ambient to 800°F, the optical thermometer reading is affected by surface roughness, surface color, and how reflective the surface is. Therefore, optical sensing of temperature is of marginal use in most lower-temperature process applications.

However, an optical pyrometer can sometimes be used to distinguish men from women. As women usually use makeup, their skin temperature will be 2°F–4°F lower than men's.

Pressure

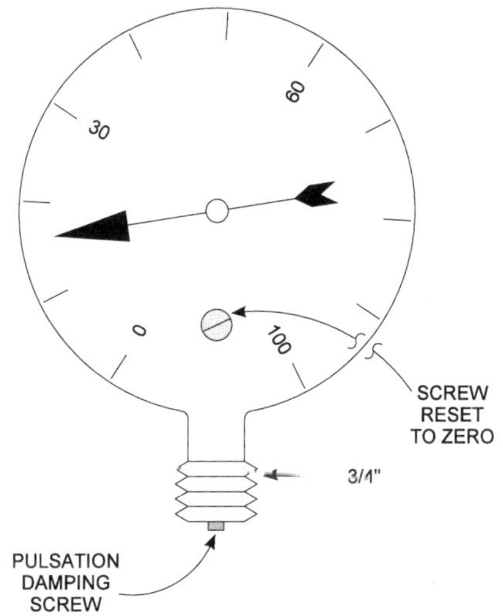

Fig. 14–2. Pressure guage

In 1960, I worked for the US Navy, recalibrating pressure gauges for an aircraft carrier. The pressure gauge screws into a pressure tap—typically, a ¾-inch screwed connection. The fluid pressure is transmitted to a thin tube in the shape of a question mark. The straight portion of the question mark is connected to the screwed inlet. The end of the rounded tube, is connected to the dial, or arrow, that twists or turns with the pressure.

This type of pressure gauge is not usually used to transmit a pressure more than a few feet. For further distances, the pressure has to be converted to an electronic signal, or to an instrument air-pressure signal.

While working for the Navy, I found that most of the pressure gauges I checked had to be recalibrated. That is, there is a tiny screw used to set the needle back to zero, at atmospheric pressure. If the gauge would still not read correctly at 100 psig, or 500 psig, I threw it away. However, most gauges will read correctly if they are rezeroed and are not plugged.

Most pressures are transmitted electronically. A "pressure transducer" converts a pressure to a milliamp output signal, in proportion to the process stream pressure.

Levels

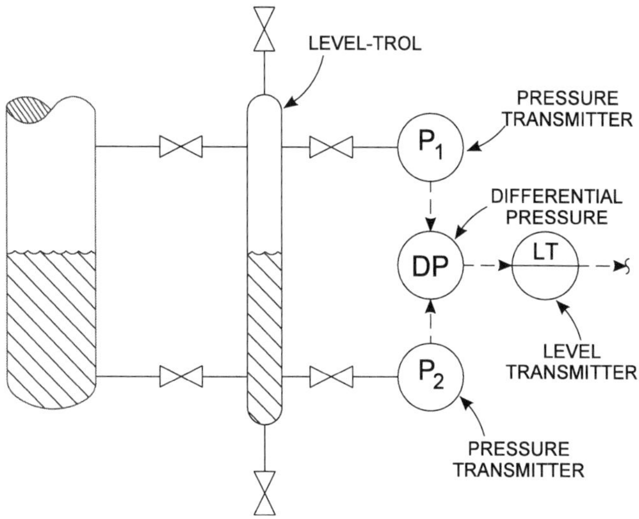

Fig. 14–3. Level indication

A hundred years ago, levels were measured by a "level-arm." The water level in your toilet's water closet has such a level-arm. That's the black float that directly controls water flow into the water closet. The old level-arm caused a pointer on the outside of the vessel to indicate the vessel's liquid level.

Fifty years ago, levels were measured by a "float chamber." This was a section of 3-inch I.D. pipe connected to the side of a vessel. Inside the pipe was a float of a slightly lower density than the liquid inside the vessel. The position of the float could be seen through the glass on the side of the float chamber.

Currently, we do not directly measure levels. We measure two pressures, using a pressure transducer, which converts a pressure into a milliamp electrical output. The dual pressure transducers span the height of the expected level in a vessel. The greater the height of the liquid above the low-pressure tap, the larger the DC electric output from the lower pressure tap on the vessel.

The vessel's level indicator converts the pressure difference (DP) between the two pressure taps to a level:

- (28) • DP (psi) ÷ (Specific Gravity) = Inches of Liquid Level.

The "28" factor is the 28 inches of water in each psi. The Specific Gravity is the actual S.G. of the fluid inside of the vessel. For example,

- Water = 1.00 S.G.
- Diesel = 0.85 S.G.
- Kerosene = 0.80 S.G.

- Gasoline = 0.70 S.G.
- Butane = 0.55 S.G.
- Propane Liquid = 0.50 S.G.

To convert from the measured DP to a level requires knowing the actual S.G. of the liquid in a vessel, at the temperature in the vessel.

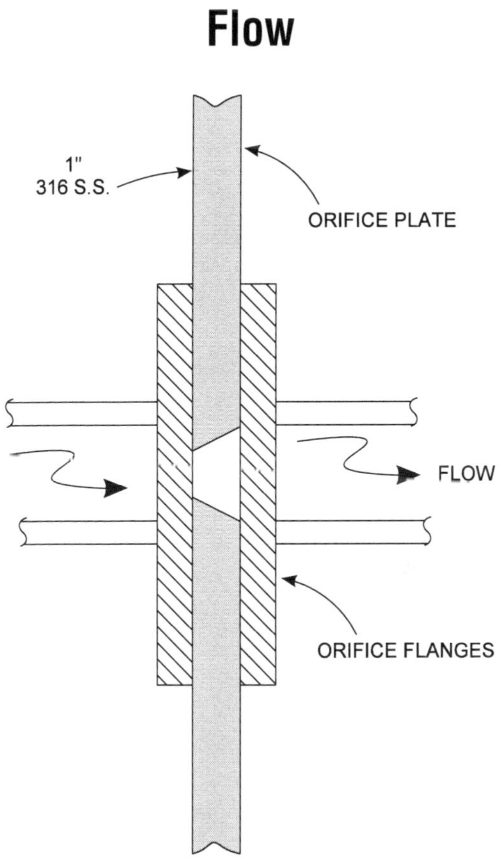

Fig. 14–4. Flow measurement

Flows are measured by passing a fluid through a restriction called an "orifice plate." It's just a plate inserted in a pipe with a hole of a known size. As a liquid or gas passes through this hole, it speeds up, or accelerates. Accelerating a flow causes the fluid's pressure to decrease. That is, we convert pressure to acceleration. The greater the pressure drop, the greater the increase in velocity of the fluid flowing through the hole in the plate.

By measuring the pressure drop (DP) through the hole, we can calculate the velocity (V) through the hole:

- DP is proportional to $(V)^2$

If I know the hole area (A) and the hole velocity, then

- Flow is proportional to (V) • (A).

One complicating factor in measuring flows is the effect of S.G. and viscosity. A bigger S.G., or a larger viscosity, will increase pressure drops through the hole in the plate. But this effect is relatively small, compared to the effect of changes in velocity through the plate (orifice plate.)

Control Valve Operation

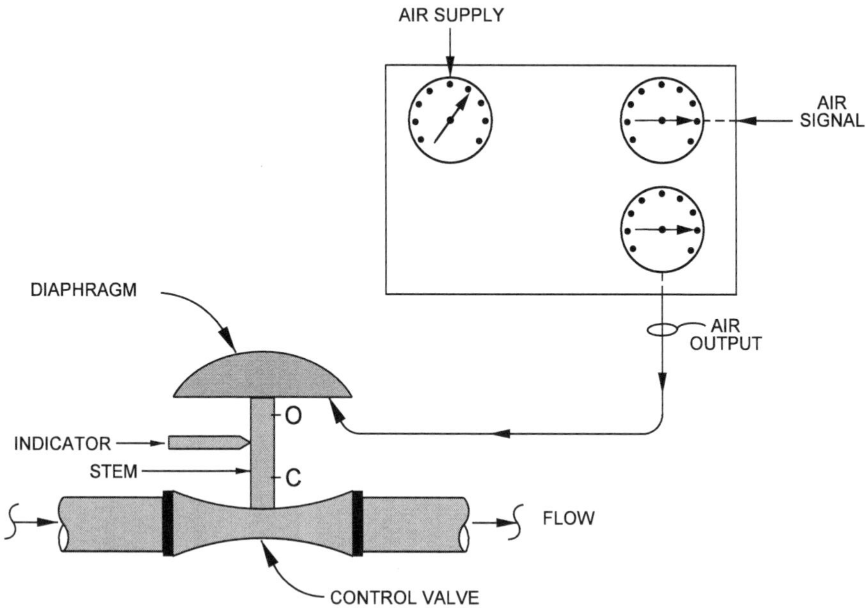

Fig. 14–5. Air to open control valve

The signal generated by the flow, or temperature, or pressure, or level indicator is typically converted into an air-pressure signal (but sometimes to an electrical signal). This air-pressure signal is sent on to a control valve, which can be either opened (Air-to-Open), or closed (Air-to-Close). The control valve then controls the level, or pressure, or flow, or temperature required by the process, in the same way that our senses of sight, smell, hearing, taste, and touch are used to control our bodies, based on the objectives our brains have set. Note that, currently, fiber optic cables are commonly used to transmit process measurements.

Process Control with On-Stream Analyzers

I have used—and have seen used—a wide variety of analyzers to control process parameters such as reflux rates, reboiler steam, and product draw-off rates:

- Colorimeter to draw off gas oil from a crude unit
- Gas chromatograph to control ethane on a de-ethanizer reboiler bottoms
- Viscometer to control stripping steam flow on a paving grade asphalt vacuum tower
- End-point on a vacuum tower gas oil draw-off flow
- Spectrometer on an aromatic (Xylene) splitter in a chemical plant
- Sulfuric acid strength, on an alkylation unit (i.e., spent acid)

In general, on-stream analyzers often are expensive, require a lot of maintenance, and have excessive lag times for signal transmission of the parameters being measured.

Instrument Air

Process units are, in general, controlled by air signals. This air, called "instrument air," is provided by an air compressor, operated by the utilities division. Air opens and closes control valves to control pressure, levels, and temperatures.

If something happens to the instrument-air system, the plant will shut down. The main problem with instrument-air reliability is moisture.

There is an "instrument air dryer" that extracts moisture from air supply to a process unit. Often, especially in warmer climates, these dryers are not maintained, as moisture in the instrument air is not a big problem.

But! When a freeze occurs, due to low-ambient temperatures, instrument airflow ceases—and the plant shuts down.

I learned the importance of maintaining my instrument air dryers in February 1976, in Texas City, on my sulfuric acid alkylation unit. I've never made that mistake again.

Lessons Learned

I have written a separate text on this subject: *Troubleshooting Process Plant Control*. Likely, 50% of all plant problems are a consequence of control problems, such as the sensing element is malfunctioning, or the control valve itself is sticking, or operating in a manner the reverse of which is intended.

15

Dangers for Automated Control

Introduction

Getting caught up in a positive feedback loop—especially in fired heaters—is an example of the dangers of automated control. Operators must understand how their controls work and be ready to switch from auto to manual. All control systems are subject to a wide range of malfunctions, and operators and unit engineers must be able to identify when this occurs.

Purpose

When a distillation tower floods, the tower-top temperature may be suppressed as cold reflux backs up the tower. Or the tower-top temperature may suddenly go up, as the cool reflux is displaced by the hot vapor from the reboiler. Either way, it illustrates the importance of the operators' understanding how their controls work under non-steady-state conditions, even when the control system itself is performing properly.

Refinery Process Controls

I worked for American Oil in operations, design, and technical management. A common example of failed process control was the raffinate splitter explosion in Texas City. This tower was designed in the early 1970s.

The operators were tracking the splitter bottoms level (Fig. 15–1). This level was measured by differential pressure. This is the modern method of level measurement, as compared to the ancient, archaic, and obsolete method of the "internal float arm."

The two pressure transmitters generate a milliamp electrical output. The difference between the dual readings (DP) is then converted to a level, based on a density (S.G.) assumed by the instrument technician at some point in time for the naphtha. I call this the "calibration S.G."

However, if the actual flowing S.G. drops by 10%—perhaps because the naphtha is hotter or lighter than normal—the indicated level transmitted to the control center will not be the actual 80%, but 72%:

- 80% • (100% - 10%) = 72%.

Referring again to (Fig. 15–1), suppose the level rises above the top tap (PT_1). Then, while the pressure at the lower tap (PT_2) will increase, the pressure at the tap (PT_1) will increase at the same rate. When the real level in the vessel reaches the top tap (PT_1), the level in the control center will only read 90%. As the actual liquid level rises by 1 foot, or 10 feet, or 100 feet above PT_1 (the top tap), the indicated level in the control room will be stuck at 90%.

We call this being "tapped out"! The indicated level then draws a straight line at 90% on the control panel.

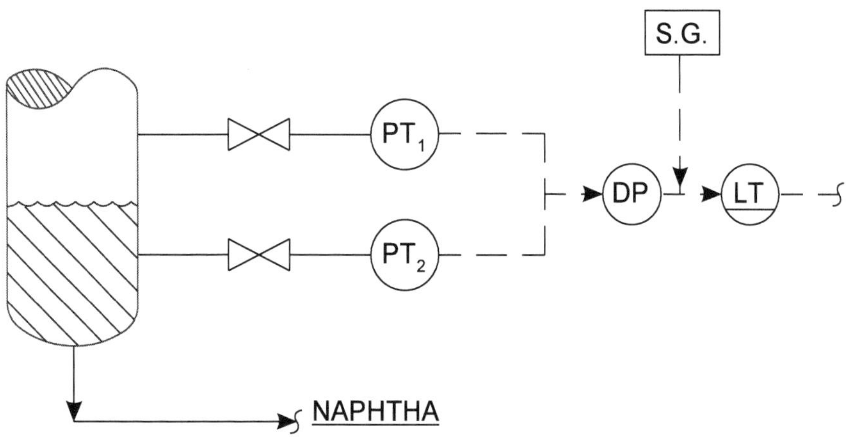

Fig. 15–1. Level measurement by differential pressure.

Manual Intervention

That absolutely straight line is the indicator to the trained operator that he or she is tapped out. The liquid level in their gauge glass is water-white, and the gauge glasses are usually scratched. Finding the meniscus is really difficult, unless you can see movement in the gauge glass.

My method is to switch the level control to manual. Then I take the pressure difference between the tower top PRC point (Fig. 15–2) and P_3, and convert the differential pressure to feet above the elevation of P_3:

- $(PRC - P_3) \cdot (2.31) \div (S.G.)$

where:

- PRC = Top pressure, psig
- P_3 = Pressure at top of upper gauge glass, psig.

The challenge is to have the operators trained to execute these steps—to run their raffinate splitter on manual—until they get their level transmitter recalibrated.

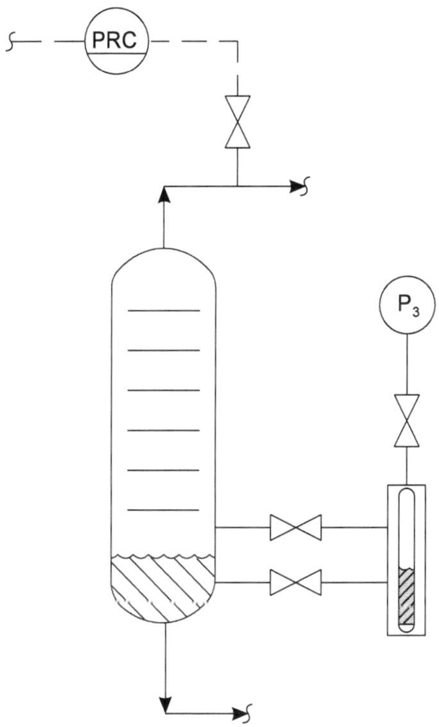

Fig. 15–2. Measure level by (PRC-P_3)•(2.31)÷(S.G.) = Feet above P-3 Guage.

Tower Pressure Control

This story dates back to April 1975, also in Texas City, where I was in charge of the #2 alkylation unit. I had 40 people working for me, and produced 26,000 BSD of high-octane gasoline. One day, as I was making my daily round on the Alky Unit, I noticed that a local pressure gauge (Fig. 15–3) was reading 450 psig.

The tower's pressure relief valve was set at 320 psig. Its mechanical design pressure is 360 psig, and its normal operating pressure is 290 psig.

I quickly checked the local gauge. It was reading correctly. What had happened?

After investigating, I determined that a corrosion inhibitor chemical we had been using had plugged both the depropanizer relief valve and the pressure-sensing tap.

Tower pressure was controlled by adding steam to the reboiler, based upon the concept that "Heat Makes Pressure." However, note from (Fig. 15–3) that both the pressure control signal to the steam supply of the reboiler and the high-pressure alarm (HPA) were connected to the same pressure point of the tower. A recipe for disaster.

I immediately switched the steam flow to the reboiler to manual, using the local pressure gauge as a temporary guide. Then I connected the pressure alarm to a different point on the tower than the pressure control transmitter (PRC).

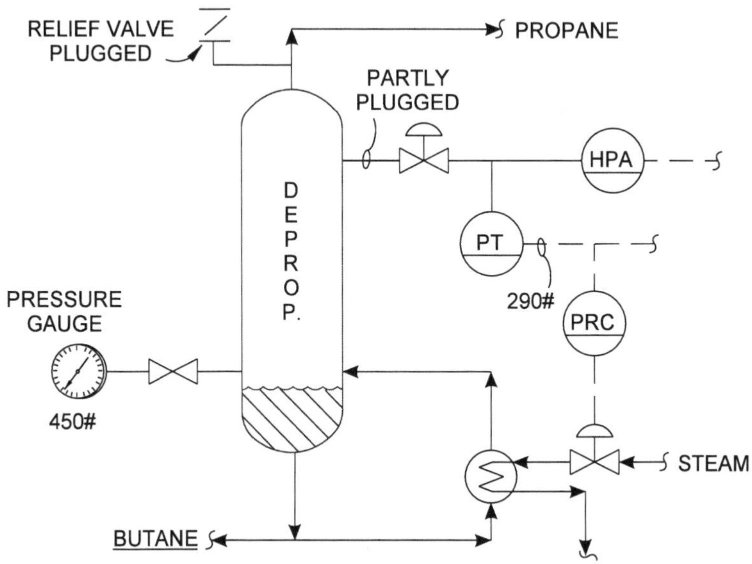

Fig. 15–3. Partly plugged pressure tap causes tower to over-pressure.

Temperature Control

During the 1980 strike in Texas City, I was working as the chief operator on the SRU (sulfur recovery plant). I was starting up the unit myself one evening, after a long shutdown. The thermal reactor had been retubed, and zirconium silicate refractory inserts had been installed to protect the tube inlets from excessive temperature. These inserts, called "ferrules," had a temperature rating of 3,000°F. Normal thermal reactor temperatures were 2,100°F–2,300°F. This temperature was monitored by a thermowell and thermocouple inserted through the 12-inch alumina refractory bricks.

During the startup, I was assisted by a young engineer on loan from the R&D department in Chicago. After a few hours of slowly heating the thermal reactor, my youthful assistant began to complain.

"Mr. Lieberman, those ferrules are glowing a yellow color. According to the color chart that we used in my combustion engineering class in college, those tube inlets are approaching 3,000°F!"

"Look here," I said. "I don't need you to instruct me! I designed this sulfur plant myself. The local TI is still quite cool at 1,800°F. Turn up the gas. I know what I'm doing."

But the thermowell (Fig. 15–4) for the local TI had only been inserted halfway (i.e., 6 inches) into the 12-inch refractory during the unit turnaround. Hence, it was reacting to a temperature halfway between ambient and the actual thermowell reactor temperature of 3,000°F.

As the heat-up progressed, I could not help but notice the ferrules inserted at the tube inlets seemed to be MELTING and restricting the small 1½-inch tube inlets.

And then, to confirm my worst fears, I noticed that the thermal reactor inlet pressure was creeping up. Apparently, the overheated thermal reactor inlet tube-sheet was restricting the flow of the flue gas due to the partly melted ferrules.

I should have been guided by my senses, not by some temperature indicator that I did not verify. I should have disconnected the remote temperature indicator, until I could verify its accuracy.

Mr. Durland, the plant manager, held me personally responsible for the sulfur recovery plant's being offline for the next two weeks. I'm much smarter now. Whenever an instrument, or a control loop, contradicts the observed plant data, I switch from auto to the manual mode of control.[III]

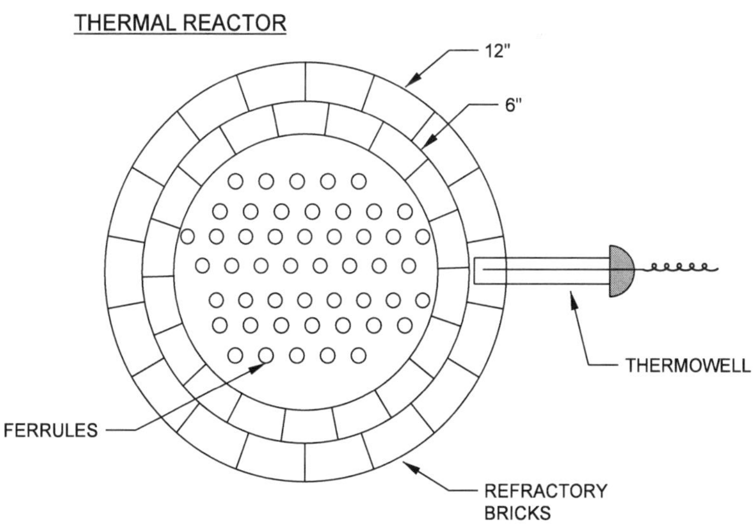

Fig. 15–4. Partly inserted thermowell causes false low temperature reading in sulfur plant start-up.

III. Lieberman, N.P. *Troubleshooting Process Plant Control*, 2nd ed., (John Wiley Publications: 2018).

Lessons Learned

Often, the control panel operator is hesitant to switch from auto to manual, even when observing that some variable is trending in the wrong direction. This, in my experience, is a consequence of poor operator training. The operator fears that, if he or she switches to the manual mode of control, the unit will spin out of control, and they will not know how to restore stability.

Most refinery and petrochemical safety incidents are, to some extent, related to controls that perform in an unintended mode. If the operators and engineers actually understand how their controls work, the unintended consequences of controls' magnifying, rather than correcting, a malfunction might oftentimes be corrected. The most common problem that I have encountered is level control in distillation towers.

16

Heat Exchangers

Introduction

Shell and tube heat exchangers are the most prevalent type of process equipment in most plants. The amount of heat exchanged in these exchangers normally exceeds the amount of heat transfer in both fired heaters and air coolers.

Hot fluid is cooled by a cool fluid via indirect contact through a small (¾-inch or 1-inch) tube. Most often, the cooler fluid is placed on the tube side of the exchanger. The hotter fluid is placed on the shell side. More frequently, the dirtier, higher-viscosity fluid is placed on the tube side, and the cleaner, less-viscous flow is placed on the shell side.

Purpose

The design of heat exchangers includes determining the following:

- Which fluid will go on which side—shell or tube?
- Number of tube side passes
- Baffle spacing on the shell side
- Baffle cut on the shell side
- Size of impingement plate
- Number of shell-side seal strips.

One key to being a competent process engineer, or an effective operator, is to understand heat transfer. Below, I have summarized common heat transfer malfunctions that I have experienced:

- Internal bypassing
- Condensation—vapor binding
- Reboiling—flux limiting
- Sensible heat transfer—low velocity
- Condensation—condensate backup
- Shell-side heat exchange—lack of vortex shedding

- Tube-side heat transfer—low Reynolds number
- Effect of surface roughness
- Fouling

Internal Bypassing

At an asphalt plant in Alabama, my client could not meet the road asphalt paving grade penetration specification, because of poor performance in their vacuum condenser. I measured the water flow in the condenser's tubes with an ultrasonic flow meter. Then, I calculated the tube side, the water DP, and compared it to my field-measured DP. My calculated DP was four times greater than my measured DP. My worst fear was confirmed. We found a leak between the channel head pass partition baffle (Fig. 16–1) and the channel head tube sheet, which permitted water to internally bypass the tubes inside the channel head.

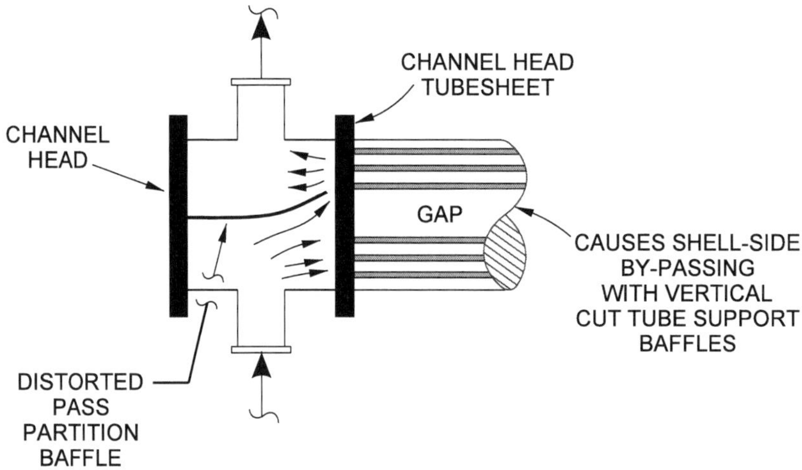

Fig. 16–1. Leaking pass partition baffle.

Since pressure drop varies with velocity squared, DP $\approx (V)^2$. Half of the cooling water was bypassing the tubes through the channel head pass partition leak.

On the shell side, the usual cause of bypassing is that the maintenance guys have discarded the shell-side seal strips that, in their minds, serve no process function. But they do. These seal strips retard internal bypassing across the top of the tube bundle (Fig. 16–2) in the gap between the tubes and the shell I.D., created by the shell-side impingement plate.

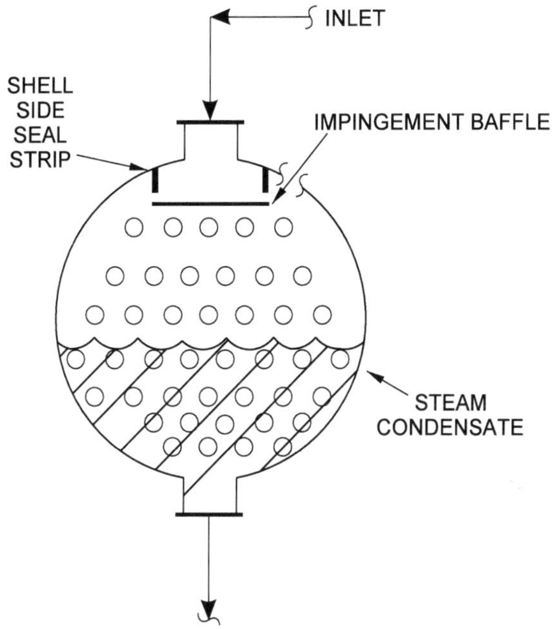

Fig. 16–2. Shell side seal strips improve heat transfer.

Vapor Binding

I have often been amazed as to how venting a little air from beneath the channel head pass partition baffle, of a steam condenser, dramatically increases a reboiler duty. Any sort of non-condensable—whether CO_2, hydrocarbon gas, or hydrogen—has the same effect. I call it vapor binding. As I've shown in (Fig. 16–3), the steam condensate runs out of the tubes, but the condensable vapors accumulate in the tubes and surround the condensing surface with a temporary fouling layer. This shields the tube surface area from contact with the steam, just like any permanent fouling deposit. A tiny amount of continuous venting of non-condensables prevents this problem (i.e., 0.5% for steam). I had a small leak of propane into the steam side in a propylene-propane splitter reboiler in Texas City, which reduced the reboiler duty by half. I corrected this by opening a vent just beneath the channel head pass partition baffle, to vent off the propane vapors.

A more common problem is the CO_2 evolved in steam boilers, when carbonates in the boiler feed water break down and leave non-condensable CO_2 in a reboiler steam supply flow. Also, hydrogen resulting from carbonic acid corrosion, has the same detrimental effect.

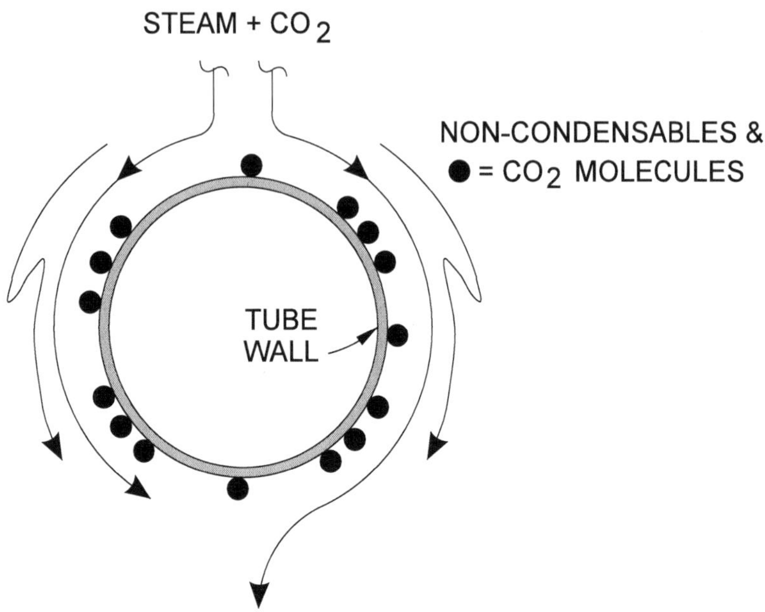

Fig. 16–3. Vapor binding reduces heat transfer.

Flux Limits on Reboilers

I once had a reboiler used in a propane-propylene splitter. It was working okay, but it was old and about 15% of the tubes had been plugged over the last 20 years. So, I decided to purchase a new tube bundle identical to the existing bundle. When I put the reboiler back into service, I was limited to 50% of the reboiler's former capacity. Quite a nasty surprise!

The reboiler used 30 psig steam. So, I tried using 100 psig, which I had connected to the reboiler at considerable expense, thinking that the hotter steam would transfer more heat. But this only made the problem worse.

Dick Nutt, an older engineer, explained the problem to me.

"Norm, get a new frying pan. Real smooth. Get it real hot. Now, gently drop a tablespoon of water on the hot pan. It ain't gonna boil away like you might think. The layer of steam that's vaporized from the water don't let the water touch the pan. If you want the water to boil away fast, shut off the burner and let the pan cool off some. Go home and try it."

So I did, but the water flashed off from the pan right away.

"Dick, it didn't work like you said. The water boiled off right away."

"Did you use an old pan or a new pan?"

"I don't have any new pans. My grandma gave me this one. I guess it's kinda rough and scratched."

"It's the same with your reboiler. The smooth surface don't have any "nucleate boiling sites" where boiling can be initiated," Dick explained.

"Oh yeah. Like in high school chemistry lab, we used those boiling stones. But what should I do about my propane reboiler?"

"Norm, put the old bundle back in."

"I sure ain't gonna do that!"

"Have the bundle coated with Linde high-flux tubing," Dick suggested.

"That sounds kind of expensive, Mr. Nutt."

"Okay, Lieberman. Than just have the exterior of the tubes sandblasted, so that they have a rough exterior surface, like the old tube bundle."

Sensible Heat Transfer

You read about this stuff in school, but it didn't mean too much. Like low Reynolds number creates high film resistance to heat transfer due to laminar flow, where

- $\text{Reynolds number} = \dfrac{\text{Velocity} \bullet \text{Density} \bullet \text{Tube diameter}}{\text{Viscosity}}$

When the Reynolds number falls below 2,100, heat transfer efficiency is supposed to fall off really fast, due to laminar flow. I found out about this at the Good Hope refinery in Louisiana. We had designed a heavy fuel oil cooler with water on the shell side and the high-viscosity fuel oil in the tubes. And it didn't work! The fuel oil product was running down to our product tank at 250°F. We blew the roof off the tank when the water in the bottom of the tank boiled. The roof sailed across St. Charles Parish and landed in someone's backyard.

So, I switched the fuel oil onto the shell side and put the cooling water in the tubes. The observed heat transfer coefficient increased by a factor of three.

Why? The velocity of the fuel oil was still low. And its viscosity was still really high.

The answer is "vortex shedding." You can see this in New Orleans as the Mississippi River flows past the French Quarter. The water flows across the old dock pilings (at about 4 mph) and creates swirls and turbulence. As long as the cross-flow velocity passing perpendicularly past the tubes exceeds 3 feet per second, the laminar film, heat transfer resistance, is largely destroyed.

Condensation and Condensate Backup

When a pound of steam condenses at atmospheric pressure and 212°F, one thousand BTUs are released. When the resulting condensate cools from 212°F to 112°F, only one hundred BTUs are released. Subcooling condensate does not release much heat. Also, it's about ten times more difficult to transfer heat to a metal surface from a liquid, than from a condensing vapor.

I can likely transfer a hundred times more heat by contacting saturated steam against a tube's metal surface than by contacting that metal surface with stagnant hot water. If I have a heat exchanger, as shown in (Fig. 16–2), those tubes that are covered by the condensate (water) can be considered to be doing very little to heat up the fluid on the tube side. This problem of condensate backup is equally valid for condensing propane and butane as it is for steam.

My mother was familiar with this problem when the radiator in our old apartment house in New York would get cold. Mom would bang on the condensate drain line with a hammer.

"Mom," I would ask, "why are you doing that?"

And Mom would say, "I'm trying to bring up the heat."

I suppose she meant to say, "I am trying to dislodge the hardness deposits from the condensate drain line, so as to expose more of the radiator's surface area to steam and retard the condensate backup, which is reducing the surface area of the radiator exposed to steam."

Operators react to condensate backup by draining the condensate to the sewer, which accounts for much of the poor steam condensate recovery in most process plants.

Blowing the Condensate Seal

When I was 19, I used to worry a lot about my girlfriend problems. Either not having a girlfriend or having multiple girlfriends simultaneously. It's the same with condensate backup. Too much condensate backup wastes heat transfer surface area. But no condensate backup is bad. It causes steam to blow through the steam side outlet.

The steam trap is intended to prevent this. The trap will open to allow water to pass, but closes to prevent steam from blowing through. When steam blows through the outlet nozzle, the operators say, "we have blown the condensate seal," and their reboiler duty will suddenly drop by around half, even though the steam flow to the reboiler has not diminished.

High velocity promotes good sensible heat transfer. But high velocity retards heat transfer from a condensing vapor onto a metal surface.

If I encounter a reboiler that is working poorly, I will perform the following test:

- Monitor the reboiler process side outlet temperature.
- Partially close the steam condensate outlet valve.
- If the process outlet gets hotter, then I have stopped blowing the condensate seal—which is good.
- If the process outlet gets colder, I have been suffering from condensate backup and need to increase the condensate (water) flow from my reboiler.

Incidentally, steam traps are not reliable. Better to use a condensate pot, and pump the condensate out of the pot through the pot's level-control valve.

Most often, plant operators manually drain condensate to the sewer at a rate that maximizes their reboiler duty. The purpose of the condensate pot and pump is to prevent this loss of valuable steam condensate to the sewer.

Vapor Lock

In my earlier years, I would design condensate from a 150 psig steam reboiler to drain through a 4-inch, 500-foot-long condensate pipe into a condensate flash drum that operated at 10 psig. The operators would then find it necessary to drain the condensate to the local ditch.

As the condensate flowed through the 4-inch pipe, it experienced a small pressure drop. The reduction in pressure caused the condensate to flash or partially vaporize. The density of the evolved steam is about 0.25 lbs per ft^3. The density of hot water is 60 lbs/ft^3. Therefore, one cubic foot of water flashing to steam expands to 240 ft^3. This chokes off the flow and pushes the condensate level up in the reboiler, which greatly reduces heat transfer rates.

The operators avoid the problem by diverting the condensate to the ditch. I now pump, rather than pressure out, the condensate to the condensate flash drum.

Tube-Side Sensible Heat Transfer

I like to keep the tube-side flowing velocity high. Anything less than 3 feet per second is too low. I design for 6 to 10 feet per second, for two reasons:

- To stay away from laminar flow due to low Reynolds number.
- Reynolds numbers are caused by high viscosity and/or low velocity.
- Fouling.

If viscosity is less than 4 to 8 centipoise (or centistokes), I don't worry about it. But for cool crude oil from Venezuela, where viscosity can be 40 to 80 centipoise, I worry about laminar flow. Not because I have read about the problem, but from experience with low-heat transfer rates on crude unit preheat exchangers running Venezuelan crude oil. High velocity will offset high viscosity.

Here are the problems with a high velocity:

Pressure drop varies with:

- DP proportional to $(V^2) \cdot (L)$

where:

- DP = Pressure drop
- V = Velocity
- L = Flow path length.

Any velocity much above 15 feet per second will promote erosion. Converting an existing tube bundle from two tube-side passes to four passes will double velocity, which is good.

But pressure drop will increase by a factor of eight:

- $DP = (2)^2 \cdot 2 = 8$.

The $(2)^2$ accounts for the velocity effect and the second "2" is because the flow path length has also doubled.

Tube Roughness

Which is better for heat transfer: a super-smooth stainless-steel tube or a rough, pitted carbon-steel tube? I used to worry about this a lot when I worked for American Oil in the 1960s. Now I realize it depends on the circumstances. If you are vaporizing Freon or boiling clean propane (i.e., pure latent heat transfer), then to avoid a normal flux limitation (12,000 BTU/HR/ft^2), a rough, pitted surface can permit a heat flux exceeding 20,000 BTU/HR/ft^2. I learned this at American Oil's El Dorado, Arkansas, refinery in 1967, on a propane refrigeration loop.

If fouling is the predominant heat transfer resistance, then smooth, shiny 316 stainless steel is the best choice. We retubed our fluid cat cracker tube bundles in American Oil's Whiting refinery with stainless to replace carbon-steel tubes and maintained excellent heat transfer coefficients. But that was all sensible heat transfer between liquid, and was not for our reboilers.

Incidentally, don't put 316 stainless-steel tubes in contact with carbon-steel tube support baffles—galvanic corrosion will destroy the baffles if water is present.

Internal Bypassing

Let's say I have an exchanger with a center baffle in the channel head (Fig. 16–1). Note that there is a gap in the center of the tube bundle shell-side between the tubes. What is going to prevent the shell-side liquid from flowing through this gap? Actually, if the shell-side baffles are cut horizontally, there is no problem. But, if the baffles are cut vertically, then the shell-side flow can bypass the tubes through this gap. I am always concerned about internal shell-side bypassing in heat exchangers.

Luckily, there is a standard solution to this problem. It's called "dummy tubes." They are used to block off the shell-side bypass area. They are inserted inside the tube bundle, but are not attached to either the channel head or floating head tube sheets.

To retard bypassing around the periphery of the bundle, I use seal strips inserted into slots cut into the tube support baffles (Fig. 16–2). Computer simulations do a nice job of predicting the benefits for these seal strip applications.

Safety Note

Do not connect a high-pressure alarm to the same connection as the pressure transmitter used for pressure control. If the connection plugs, then the parameter controlling the vessel pressure (i.e., reboiler steam, condenser hot-vapor bypass) will open fully, as the indicated pressure drops. But no alarm will sound. This happened to me on my alky depropanizer in Texas City. The relief valve stuck, and the vessel, with a maximum allowable working pressure of 330 psig, went to 450 psig before it was noticed by an outside operator, who was fortunately looking at a local pressure gauge.

Lessons Learned

The important lesson to be learned from this chapter, is do not leave the following items to the exchanger equipment vendor:

- Number of tube-side passes
- Baffle spacing on the shell side
- Baffle cut on the shell side
- Size of impingement plate
- Number of shell side seal strips.

Use your HTRI (Heat Transfer Research, Incorporated) computer simulation as a guide to specify these parameters yourself.

17

Cooling Towers: Minimizing the Water Return Temperature

Introduction

The cooling medium for most process plants comes from either a natural draft or an induced draft (like using a fan), to cool the circulating water by evaporation. The sensible heat of the water is converted into latent heat of evaporation by contact with a relatively dry airflow.

Purpose

The principal purpose of the cooling tower is to cool the circulating water, as close as possible, to its water dew-point temperature, which is a function of the air temperature and also the relative humidity of the air. This is called the "approach to wet bulb temperature." An approach to wet bulb of 3°F–4°F is excellent cooling tower performance. An approach to wet bulb of 10°F–12°F is very poor performance of the cooling tower. Design is typically 6°F–8°F for new towers.

Mankind has selected water as cooling fluid. As a supervisor of a refinery process unit in Texas City, where we produced a million gallons a day of high-octane gasoline, most of my challenges were related to cooling water circulation.

Upon arriving in Texas City in 1974, I immediately became aware that the unit suffered from a lack of cooling capacity. The critical distillation towers were over-pressuring, due to insufficient condenser capacity. Back-pressure from the steam turbine surface condensers were causing turbines to run slow. Especially during the July heat, the poor performance of the cooling water tower had to be corrected.

Cooling Towers

The most prominent feature of many industrial facilities is the cooling tower. There are two sorts of cooling towers. The really tall cooling towers (Fig. 17–1) are "natural draft." The more common sort of cooling tower is the shorter "induced draft," which uses a motor-driven fan to induce airflow into the cooling tower.

The air is drawn through wooden boards on both sides of the cooling tower cells. At Texas City, I had eight of these cells. Each had its own fan. The fans drew the air into the cell through the side-boards. At first, the air flowed horizontally into the cell before turning upwards. Inside the cell, there were stacks of wooden slats to promote air-water contact.

A door was located in the side at both ends of the string of cooling water cells. A wooden 18-inch-wide walkway led through the interior of the eight cells. I recall walking through my cooling tower cells and noting that hundreds of the wooden 2-inch-wide distribution slats were broken and dislodged. This apparently reduced the contact between the air, which was drawn directly upwards by the eight giant rotating fan blades, and the descending circulating water.

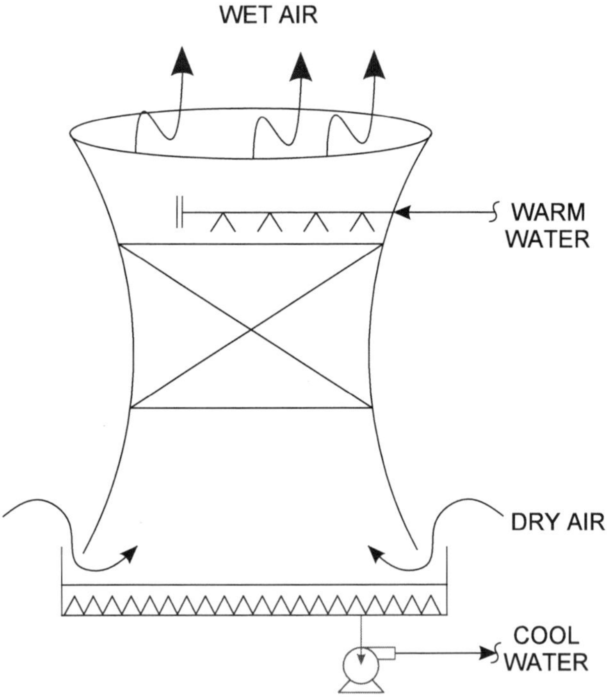

Fig. 17–1. Induced-draft cooling tower.

Distribution Decks

The warm cooling water return flows onto the top of the distribution decks. The condition of these water distribution decks is critical to efficient cooling of the circulating water. The distribution decks are located on either side of the fan. They are about 30 feet wide and 20 feet long. Each deck has hundreds of 1-inch holes, with rubber or plastic inserts in each hole. These holes have a propensity to plug. If enough holes plug, then two bad things transpire:

1. Water starts to overflow the distribution decks, at the sides of the cooling tower cells.
2. Portions of the interior of the cooling tower fill (like the slats) dry out.

Both factors negatively affect air-water distribution and thus increase the cooling water return temperature to the process unit's heat exchangers and condensers.

The origin of the holes' plugging is organic sludge. The sludge is a result of hydrocarbon leaks into the circulating cooling water circuit. H_2S, NH_3, and phenols, in addition to propane, butane, gasoline, and diesel, are food sources for bacteria that create the sludge, which plugs distribution deck holes.

There are two methods to clear these holes of the organic sludge:

- *The Wrong Way.* Insert a ½-inch rod through each hole, being careful not to slip and fall off the distribution deck.
- *The Right Way.* Establish a chlorine concentration of 2–5 ppm in the circulating water for perhaps 4–8 hours. This will kill off the bacteria. The cooling tower will foam up dramatically during its "shock chlorine" treatment. It's best to use granulated chlorine, rather than chlorine cylinders, which may be hazardous.

Preventing Distribution Deck Fouling

The best way to prevent sludge formation on the water distribution decks—which is an industry-wide problem—is not by maintaining a high-chlorine residual in the circulating water. This extracts the lignite content from the cooling tower's wooden components (especially the wooden distribution slats) and reduces their strength. Nor is the correct way the use of chemical biocides, offered by chemical vendors. Biocides are expensive, and do not eliminate the need for shocking the cooling tower with chlorine periodically.

The correct, environmentally sound, and profitable method to stop sludge formation on the cooling tower's water distribution decks is to find and fix the hydrocarbon leaks in the system's coolers and condensers. These hydrocarbon leaks can, and have, caught fire and burnt down cooling towers.

Cooling Tower Efficiency

The procedure for judging the efficiency of a cooling tower is based on the cooling tower's water outlet temperature, as compared to the atmospheric "wet bulb temperature." The wet bulb temperature depends on the ambient temperature and the humidity. Just wrap a wet rag around the bulb of a glass thermometer and swing it around for a minute. The thermometer will now read the wet bulb temperature. Compare this to the cooling tower water outlet. This is the approach to wet bulb. The following chart represents my standard for the cooling tower performance.

Table 17–1. Cooling tower performance standard

Approach Temperature	Rating
0°F	Theoretically perfect
4°F (minus)	Really excellent
6°F	Good to okay
8°F	Marginally okay
10°F	Bad
12°F (plus)	Really Terrible

Perhaps the most common cause of a large approach to the wet bulb temperature, after bad water distribution, is the lack of airflow from the induced-draft cooling tower fan, located in the cooling tower chimney (Fig. 17–1).

Airflow

Cooling towers are typically designed with an induced-draft fan that draws relatively dry air through the sides of the cooling tower, which then flows vertically through the cooling tower's wooden slats. The water flow trickling down the slats comes into close contact with up-flowing air.

As it mixes with the water, the air, which is not yet saturated with moisture, causes a few percent of the water to evaporate. The heat to evaporate the water comes from the temperature of the water.

That is, the sensible heat of the water is converted into latent heat of evaporation of water. Evaporated water increases the humidity of the air. If the air starts out at 100% relative humidity, then the air will not be able to cool the circulating water by evaporation.

It's the flowing air, drawn into the cooling tower, that cools the circulating water. It's not the temperature of the air, but the relative humidity of the air that's important.

Calculating How Much Water Evaporates

In Texas City, in July, the cooling water return temperature from my process unit was 120°F. The cool water supply temperature, leaving the tower, was 90°F. The heat removed from each pound of water was:

- (120°F - 90°F) • (1.0 BTU/LB/°F) = 30 BTU/LB.

The latent heat needed to vaporize a single pound of water is about 1,000 BTU/LB. Therefore,

- 30 ÷ 1,000 = 3%.

It was this evaporating water that cooled the circulating water. On those days in Texas City, when ambient conditions exceeded 95°F, our cooling water supply temperature could be 85°F–89°F, if the relative humidity was very low.

Airflow Problems

There are three factors that promote good cooling of the circulating water:

- How well the air and water come into contact inside the cooling cell
- The relative humidity of the air
- The amount of flowing air

Usually, fans that generate the airflow are driven by alternating-current, fixed-speed motors, running at 3600 rpm in the US (3000 rpm in European and Asian locations). The fan speed is very much lower and can be adjusted by changing the wheel (or pulley) size on the shaft of the motor. Increasing the wheel size by 10% increases the fan speed also by 10%. But, it increases amp load on the motor by

- $(1.10)^3$ = 1.34 or 34%.

The exponent of "3" is derived from the "affinity law" for rotating equipment. Often, the amp rating of the motor (FLA point) does not permit a bigger wheel to be installed on the motor's shaft.

Reduction of the clearance between the tip of the fan blade and the interior of the chimney is an effective way to increase airflow. This is done by installation of "seal strips" in the interior of the chimney. The strips are available from the supplier of the "Fin-Fan Air Coolers."

A simpler method to increase airflow is to increase the fan blade pitch or angle. The maximum is 22½°. Going from 15° to 20° will provide a few more percent airflow. Don't expect too much from this change.

Cycles of Concentration

Typically, 3% of circulating water evaporates with each pass of the cooling water through the tower. This might suggest that only 3% of new water has to be added to the cooling tower, based on the water circulation rate.

However, as the water evaporates, hardness deposits in the circulating water increase. This increases the tendency of the cooling water to foul heat exchanger surfaces.

The rule of thumb for the allowable increase in hardness deposits is a factor of six. To prevent hardness deposit concentrations from exceeding this six factor, cooling towers are equipped with a "blow-down" drain to the sewer. However, in my experience, I have never found it necessary to intentionally discard water from a cooling tower to control the accumulation of the hardness deposits. On my Alky unit in Texas City, there were sufficient unintentional losses of water to control buildup of hardness deposits.

Water makeup is, of course, required. This is done by level control in the sump that the circulation pumps draw upon. The makeup water has sediment removed, but not dissolved minerals.

The sediment that accumulates in heat exchangers on the cooling-water side, according to a lab analysis, was not silt from the cooling tower's makeup water. It was 98% organic material from biological growth in the cooling water, due to hydrocarbon leaks in the water circulation system. The best way to prevent sludge accumulating in the bottom of cooling towers is maintaining a chlorine residual of about 1 ppm—and fix hydrocarbon leaks into the cooling-water circulation.

Cooling-Water Return Temperature

Depending on the concentration and type of hardness deposits, the maximum water outlet temperature from individual coolers should be in the range of 125°F to 135°F. Above these temperatures, the rate of fouling due to water deposits will increase exponentially. Experienced operators will reduce a distillation tower's reboiler duty or feed rate, to avoid an excessive cooling water outlet temperature from a condenser.

This critical temperature is a function of the plant's cooling water chemical treatment program, the cycles of concentration, and the source of makeup water. The normal design temperature limit for the cooling-water outlet is 120°F. Without actual analysis, I would take corrective action if the cooling-water outlet, during a hot summer's day, exceeded 130°F.

Once a tube starts to foul, water flow is restricted. This does not increase water pressure drop, but does reduce water flow, with the following consequences:

- Increased water outlet temperature
- Increased rate of fouling
- Unchanged water flow pressure drop
- Increased rate of water fouling, due to fouling restricts water flow and raises temperature more. The problem then feeds upon itself, (i.e., a positive feedback loop is created).

Lessons Learned

Maximizing the airflow through the cooling tower, as well as good water distribution across the distribution decks (or through the spray nozzles), are the key elements in minimizing the cooling-water return temperature. Suppressing the formation of sludge, principally due to hydrocarbon leaks, and maintenance of an adequate chlorine level in the circulating water will aid in a good approach to the wet bulb temperature and minimizing of the cooling-water return temperature. Providing a method to extract sludge from the circulation cooling water is a critical method to reduce exchanger fouling on the water side of the process-water-cooled heat exchangers.

18

Cooling-Water Circuits: Maximizing Circulation Rates

Introduction

Lack of cooling-water flow to the distillation tower's overhead condensers is the bottleneck on many fractionators. Without adequate condenser capacity, tower pressure becomes excessive and the reboiler duty must be reduced, which reduces the reflux rate, as the reflux comes from the reboiler, as previously explained.

Purpose

Understanding reduction in the flow of the circulating cooling water is key to identifying the causes of—and finding cures for—sludge formation in the circulating water. One common problem is carbonate deposits in the cooling-water tubes and in the circulating cooling-water lines.

The most common problem associated with low cooling-water flow is excessive pressure drop in the cooling-water return header. This line should be cleaned every turnaround. Back-flushing with a nitrogen purge to promote turbulence is an effective aid in increasing water flow.

Eliminating air leaks at the suction of the cooling-water circulation pumps (which are normally elevated above the cooling-water basin level) will substantially improve water flow. Sludge accumulation is best suppressed by eliminating hydrocarbon leaks in coolers.

Cooling-water circulation is often regarded as a process variable of little concern to unit operators. Also, the cooling-water return temperature is thought to be only a function of ambient conditions. Reduction in the performance of the cooling-water system is often not monitored. However, the reduction in cooling capacity as a result of reduced water flow and increased supply temperature reduces product recovery, operating efficiency, and unit capacity.

Cooling-Water Systems

Cooling-water supply is from a central cooling tower (Fig. 18–1). Water is distributed over wooden slats or other types of contact device. Air is drawn up through the slats through a chimney. When a fan is used, it's called induced draft. If a chimney is used, it's called natural draft. The water is cooled by evaporation. For this to be possible, ambient air needs to be less than 100% humidity—that is, below its dew-point temperature. Here is what is required to cool 100,000 lbs of water (or 200 GPM) from 100°F to 80°F:

- $$\frac{(100{,}00 \text{ LB}) \bullet (100°\text{F} - 80°\text{F})\ (1.0 \text{ BTU} / °\text{F})}{1{,}000 \text{ BTU/LB}} = 200 \text{ pounds of water evaporated}$$

where:

1.0 BTU/°F/LB is the specific heat of water, and 1,000 BTU/LB is the latent heat of vaporization of water per pound. For the water to evaporate, the air must be below its "wet bulb temperature" (see previous chapter for explanation).

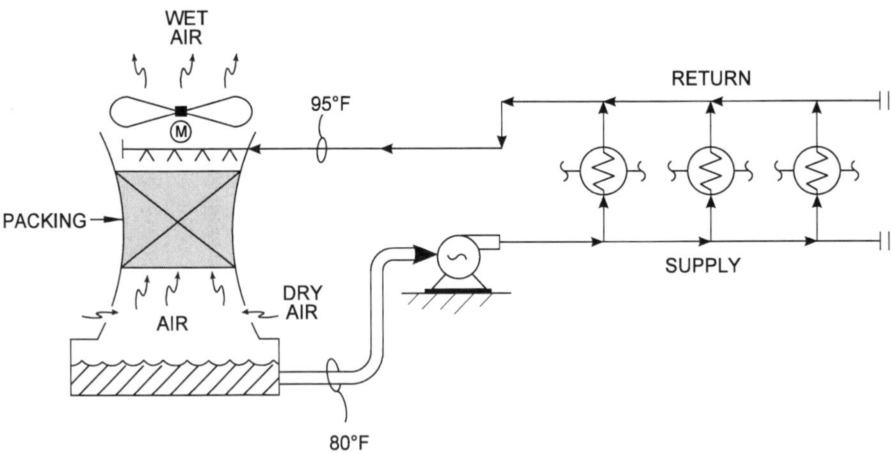

Fig. 18–1. Induced-draft cooling tower

Water Pressure

Both the supply and return pressure of the cooling water appear to the plant operator to be constant—with the supply pressure being the cooling tower pump discharge pressure (typically 60 psig) and the return pressure being that required to lift the water to the top of the cooling tower (typically, 20 psig or 45 feet). The difference between the 60 psig and the 20 psig (being 40 psig) is not the total pressure drop available to circulate the cooling water through the unit's water coolers. You must also take into account the pressure loss in the supply and return water circulation lines and valves.

Time, Temperature, and pH

"Norm, we require a computer analysis of our naphtha reformer. Are you interested in providing a quotation for this project? Our lack of cooling capacity on the reactor effluent is overloading our recycle gas centrifugal compressor motor, due to excessive molecular weight of the recycled hydrogen," Jane asked.

"Jane, I suggest $50,000."

"Can we then base the contract on a 'not to exceed $50,000' basis?"

"Okay. But our computer correlations are proprietary, and while we will tabulate results, we cannot share our correlation."

"I quite agree. Only results need to be reported on the spreadsheets."

Pressure Drop Survey

(Fig. 18–2) shows the results of the field pressure drop survey of the plant's cooling-water circuit:

- Pump suction = 6-inch Hg vacuum
- Pump discharge = 67 psig
- Cooling-water pressure at exchanger inlets = 58 psig (at grade)
- Cooling-water pressure at exchanger outlets = 48 psig (10 ft. above grade)
- Cooling-water pressure return at the bottom of the tower = 12 psig

The 6-inch Hg pump suction pressure is equivalent to a vacuum of -3 psig. That is, each 2 inches of mercury vacuum equals 1 psi of negative pressure. Thus, the pump was developing:

- 3 + 67 = 70 psi of pressure.

The line loss from the pump discharge to the exchanger inlets on the reformer unit was:

- 67 psig – 58 psig = 9 psi.

The pressure drop through the process coolers was:

- 58 psig – 48 psig = 10 psi.

Correcting for the 10 feet of elevation (2½ ft. of water = 1 psi), the pressure loss due to friction in the coolers was:

- 10 psi – (10 ft/2.5 ft) = 6 psi.

And the pressure drop through the return header was:

- 48 psig – 12 psig = 36 psig.

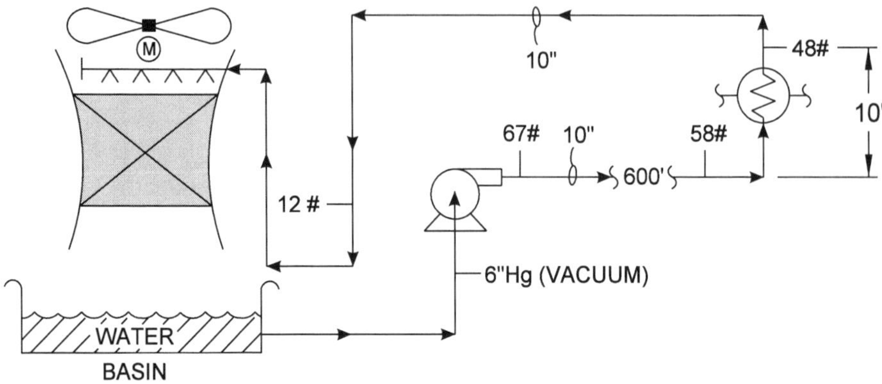

Fig. 18–2. Pressure drop survey

Frictional Loss in Water Piping

I use a "Doppler flow meter" to externally measure velocity in cooling-water lines. At the cooling-water pump discharge, I measured a velocity of 8 feet per second. Line size was 10 inches. Using a relationship for pressure drops in clean piping,

- $DP = \frac{4}{I.D.} \bullet \frac{(DEN)}{62} \bullet \frac{(V)^2}{28}$

where:

- DP = Pressure drop in 100 equivalent feet, psi
- DEN = Density in lbs/ft^3
- V = Velocity, feet per second
- I.D. = Pipe inner diameter, inches.

I calculated that for the 600 feet of mainly straight piping, the pressure drop in the cooled water supply line should be 7 psi, as compared to the 9 psi observed. As the cooling-water return line ran parallel to the supply line, and flow and I.D. were identical, I expected the return line pressure drop would also be about 7 to 9 psi. But it was the 36 psi noted in (Fig. 18–2)! Something was very wrong?

Hardness Deposits

"Norm, when we open our exchanges to clean out deposits from our water coolers, our supply line has an inch or less of hard deposits. Looks grayish-white," Travis, the outside operator, noted.

Well, since pressure drop varies with pipe diameter, as follows:

- DP is proportional to $(I.D.)^5$

where:

- DP = Pressure drop, psi
- I.D. = Pipe diameter, inches,

then for a 10-inch pipe with an average of an inch of deposits (i.e., 8 inches of pipe I.D., average), DP would increase by:

- $[10" \div 8"]^5 = 2.0$.

That is, the the one-inch deposits would cause the pressure drop in the water supply line to double, from about 7 psi to 14 psi. But how about the 36-psi pressure drop in the warm cooling-water return line? Also, what had caused the deposits?

Water pH Control

I checked the circulating water pH. It was 6½.

"Norm, our old manager tried to save on cooling tower chemicals. So we used to run 7½ plus pH. Our new manager told us to follow our chemical treatment guidelines from the water treatment vendor, which says to hold at 6½ pH by addition of sulfuric acid. Our exchanger doesn't foul up so fast since then."

My equation for frictional loss in piping assumes normal pipe roughness and not that of the rougher hardness deposits. That was also part of the high cooling-water return pressure-drop problem. The 6½ pH target has to be adhered to carefully. Using extra acid to produce a lower pH will greatly accelerate corrosion rates in the carbon steel cooling-water pipes. Water lines can be chemically cleaned with a "chelating solution."

A Localized Restriction

I did a second pressure survey (Fig. 18–3). The 10-inch line, just before it returned to the cooling tower, dipped underneath a gravel road. On either side of the road, I could measure the water pressure. The observed DP was (26 – 12) = 14 psi, compared to the calculated DP of only one psi! The next day, I had the road excavated and found a 10-inch gate valve, mostly closed, buried underneath the road, which I could not open.

Why this valve was installed, no one knew. Or why it was mostly closed was also a mystery. However, there was a third problem. Having calculated water circulation rates from the observed water velocity at the pump discharge and the 10-inch line, I realized the pump was running well below the manufacturer's flow vs. head performance curve.

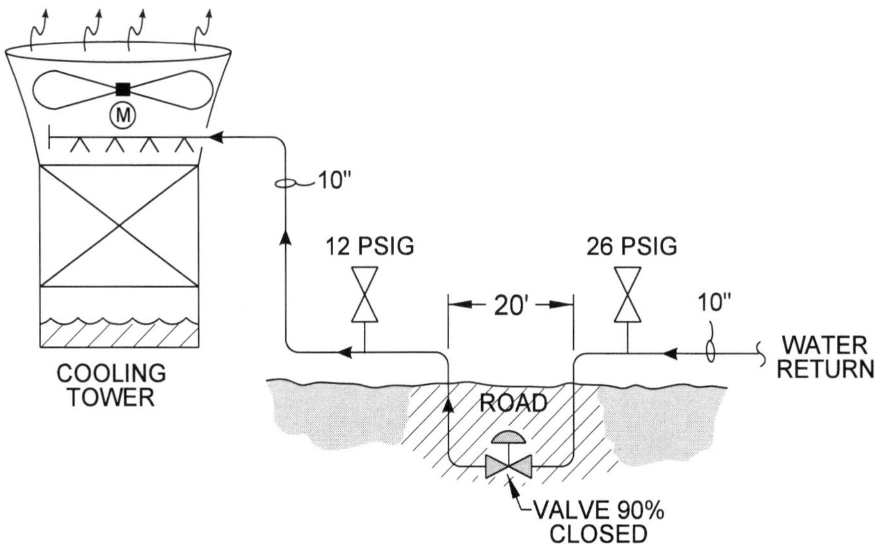

Fig. 18–3. A buried valve partially closed restricted cooling water circulation.

Air Leaks in Suction

"Jane, I want to show you something."

"What's that, Norm?"

"I want to show you what happens when I shut down the spare cooling-water pump."

I left the "A&B" pumps running and pushed the stop button on "C." A few small jets of water blew out of the "C" pump's suction flange and also from several small holes in the suction line.

"What's the point, Norm?" Jane asked. "It's only water. And the leaks are small. I can't see any safety problems."

"The pump suction is under a slight vacuum, because the pumps are elevated above grade (see Fig. 18–1, 18–2) by about 6 feet. Air is drawn into these leaks, into the pump suction and through the pump's impeller. The air bubbles reduce the flow of water, and the pump's discharge pressure and flow," I said.

"I believe you're right. I've noticed that, when it rains hard, our cooling-water pumps' discharge pressure increases. I think that's because the rain replaces the air leakage with water."

And when we sprayed a torrent of fire water across the suction lines of the "A&B" pumps, their discharge pressure increased by several psi.

Approach to Wet Bulb

"Norm, did you ever check our approach to the cooling tower's water return temperature as compared to the wet bulb? The tower is designed at 20,000 GPM, for an 8°F approach."

"I have. You're running at 10,000 GPM, with a 13°F approach to wet bulb."

"Any suggestions to reduce this approach?"

"Water distribution may be bad. Perhaps, the wooden slats in your cooling tower have deteriorated. Check that the fan belts are not slipping. Increase the fan blade angle to 22°, and check that the blade tip-to-shroud clearance meets design specification. If the motor amps allow, run the fan faster, by putting a smaller wheel on the fan's shaft or a larger wheel on the motor shaft. This will help increase airflow, but motor amperage will increase a lot."

"Norm, I do not quite see how any of your recommendations are based on computer modeling. We agreed to pay up to $50,000 for this study. Where is the computer study?" Jane asked.

"Jane, you don't understand. In process operations, only results count. You have received $50,000 worth of results," I explained.

Safety Note

Hydrocarbon leaks in water-cooling service can cause flammable mixtures to form below the distribution decks in cooling towers. Especially when cooling butane streams, this can, and has, caused cooling towers to ignite. Check cooling-water outlets for combustible vapors perhaps once a month. The instruments operators use to issue vessel entry permits are used for this purpose. Vapors accumulate beneath distribution decks. Sludge buildup on the decks is another indication of hydrocarbon leaks into the cooling-water return.

Washing of Hydrocarbon Liquids

To remove the salts and the finely divided solids dispersed in liquid hydrocarbons such as propane, kerosene or crude oil, water is mixed into the hydrocarbons. Water-soluble salts are then extracted into the water.

To separate water and hydrocarbon liquids by gravity settling could take an hour. To speed settling to 10 minutes, an electrical field is generated inside a vessel called a "precipitator." The electric grids cause dispersed droplets of water to coalesce into larger drops, which then settle into the bottom of the "coalescer vessel."

Salty water, or brine, is drawn off the bottom of this vessel, and clean hydrocarbon liquid flows from the coalescer's top.

Lessons Learned

Cooling-water circulation problems are complex, but often ignored in most process plants. Monitor the cooling-water supply and return pressures. In many plants with an excessive pressure drop in the warm-water return line, the system has been in service for over 20 years, yet the return line has never been cleaned. A circulating chelating chemical for off-line cleaning during the unit turnaround, has been used in some plants to rectify this problem.

19

Motive Steam: Use in Turbines and Engines

Introduction

The greatest use of energy worldwide is to drive steam turbines to generate electricity. That applies to nuclear energy, coal-fired plants, and power generation from industrial fuel oil. Also, steam turbine drives are extremely common in process plants and particularly in refineries. Thermodynamics, for me, is the use of the Mollier diagram to design and monitor the performance of both topping steam turbines and turbines exhausting to a vacuum surface condenser.

Purpose

The main concepts in this chapter are the efficient use of:

- The governor speed control valve.
- The 2- or 3-horsepower valves on the steam chest.
- The downstream process control valve.

These three items should perform as a team to optimize energy efficiency. I have reduced turbine steam usage at a refinery by 30% (Texas City, 1980) by optimizing the above three variables.

Coal, nuclear energy, and fuel oil are not used to generate electricity. These sources of energy are used to boil cold water into hot steam. It's the steam that spins a turbine. As the turbine spins, it forces a coil of wire, attached to the turbine's shaft (rotor), to cut through a magnetic field created by a stationary magnet. The repeated cutting of the magnetic field of the stationary magnet causes electricity to flow from this stationary (stator) magnet into wires attached to the plus and minus sides of the stationary magnet. That's how "direct current" electric power is generated. Nikola Tesla also used this concept to invent the "alternating current" motor in the 1890s.

In a modern power plant, steam is generated at 1500 psig (100 BAR). In a process unit, there are usually two steam supplies:

- 450 psig (30 BAR)
- 150 psig (10 BAR)

Between 1974–1976, I supervised a process unit with a 10,000-horsepower butane compressor, driven with a 150 psig steam turbine. Even though the compressor was the major component of my unit, which produced 1,000,000 gal/day of gasoline, I had no idea how the steam turbine driving the compressor worked. But wisdom comes with age.

Enthalpy and Kinetic Energy

I used to think that the property of steam that caused a turbine to spin was the pressure of the steam. Quite wrong! The property of steam that causes a turbine to spin is not pressure, but velocity. When the motive steam impacts the blades on the rotor of the turbine, it's already essentially (less minor "windage" losses) at the turbine's exhaust steam pressure.

What then causes the turbine to spin with great rotational force, enough to drive a 10,000-horsepower compressor? It's the steam's:

- Kinetic energy, or
- Speed, or
- Velocity, or
- Momentum

that provides energy to spin the turbine and compressor.

But where does the steam derive its speed from? Not from the pressure of the steam, but from the steam's:

- Enthalpy, or
- Heat content, or
- Temperature, and the
- Latent heat of condensation.

Critical Velocity

The speed of sound in air is 1,100 ft/second. In different fluids with different densities than air, the speed of sound varies, but not by much.

If I drill a hole in a 400 psig steam pipe, the steam will blow out of the hole with an exit velocity in the range of 1,100 feet per second. That's it!

However, if we allow the steam to fall in pressure through a specially shaped nozzle (Fig. 19–1), then the exhaust steam nozzle exit velocity can be 12,000 feet per second (i.e., 10 times sonic velocity), or more. This gigantic amount of kinetic energy impacting the steam turbine's blades causes the turbine to spin the compressor with great rotational power.

But from where does the steam derive this kinetic energy, or speed, or momentum?

Not from the steam's pressure. Rather, from the heat, or enthalpy, or the temperature of the steam.

I've done the calculations rigorously (using a Mollier diagram) for a steam turbine driven by 150 psig saturated steam, where the turbine exhausts to a vacuum surface condenser. The motive energy to drive the turbine (assuming 100% efficiency) is derived from:

- Steam pressure: 1%
- Steam temperature: 62% (the steam cools)
- Steam latent heat: 37% (the steam partly condenses)

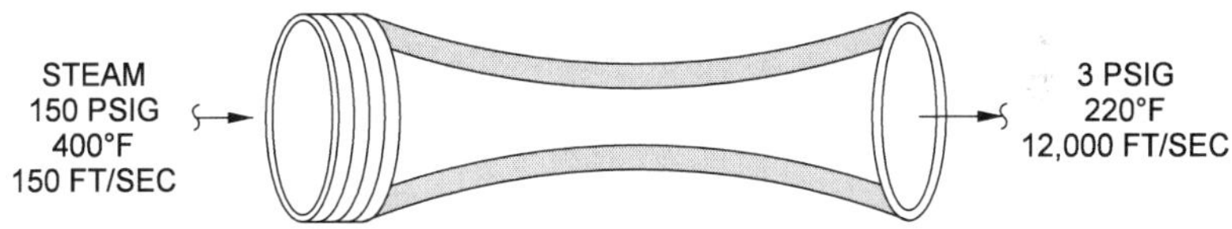

Fig. 19–1. Turbine steam nozzle

Source of Heat

Making steam requires heat. Until recently, this heat came mainly from burning coal. Lately, about 15% of electricity in the US is derived from the radioactive decay of U-235 in nuclear power plants.

Gas-fired turbines (which I used in the Laredo, Texas, natural gas fields) do not use steam. The combusted 1800°F gas impacts directly on the turbine blades. A jet engine works in the same way, except kerosene is used instead of natural gas.

In archaic, reciprocating steam engines, used to power river boats on the Mississippi River in the 1800s, the energy to drive the engines did not come from the velocity of the steam, but from its "latent heat of condensation." The energy released by the partly condensed steam pushes the piston inside the cylinder.

Wet Steam Effect on Turbines

If steam is used in a reboiler and it contains a few percent moisture, the ability of the steam to provide heat is not diminished, except in proportion to its water content. If steam is used to drive a turbine and it contains a few percent moisture, the ability of the steam to drive the turbine will be greatly diminished.

The reason for this is that the property of the steam spinning the turbine is the velocity, or the speed, of the steam. Rather like a windmill that is driven by the wind. And the velocity of the steam is not derived from the steam's pressure, but from the heat of the steam. If the steam has a few percent moisture, then as the steam flows through the "steam nozzle" (Fig. 19–1), the moisture in the steam evaporates.

For example, if steam has 5% water, then the steam would cool by about 88°F, just to evaporate this moisture content. The reduction in temperature of expanded steam, after it flows through the steam nozzles, but before it impacts the turbine blades, comes at the expense of motive steam kinetic energy, or momentum, or its velocity.

In Texas City, I would sometimes observe that my giant butane compressor would slow, even at a constant steam pressure and flow. I wondered why. Now I know! Wet steam! Likely, due to moisture carryover from an upstream boiler.

Wet steam can also be bad for a turbine, if the water droplets contain hardness deposits, carried over from the boiler. The water droplets can contain minerals, such as silicates, which solidify on the steam nozzle throat, turbine blades, and surface condenser.

In summary, avoiding boiler carryover and ambient heat losses in the steam distribution piping are good methods to preserve a steam turbine's horsepower potential.

Reciprocating Steam Engine

We learned in school that the industrial revolution in the 1800s was initiated by the invention of the steam engine. However, instructors in thermodynamics often did not understand themselves the source of energy that motivates steam engines.

Let's assume steam is coming from a 150# boiler. The steam is saturated at 360°F. The 150# steam will be pushing against a piston. The piston is driving a river boat. Here is what happens to the steam:

1. The 150# saturated steam at 360°F enters the cylinder through the open intake valve.
2. The cylinder discharge valve remains shut, as long as the intake valve is open.
3. At all times, when the intake valve is open, and the discharge valve is shut, the pressure inside the cylinder is the 150# steam supply pressure.
4. The 150# steam pushes against the piston, and this does the work of driving the paddle wheel on our river boat.
5. The steam pressure and temperature inside the cylinder is constant (because the inlet valve is 100% open, and the discharge valve is completely shut), and the steam supply is at its saturation 360°F temperature. So, the temperature of the expanding steam pushing the piston—and doing the work—is also a constant 360°F.

Here's the question: if both the pressure and temperature of the expanding steam are constant, where does the energy come from to drive our river boat? It comes from latent heat of condensation of steam. About 20% of the steam turns into water. We are converting heat into work.

As steam expands inside the cylinder, with the intake valves open and the discharge valve shut, more steam in drawn into the cylinder, to replace the 20% of steam that has condensed.

The piston, having completed its travel, being linked to the valves, causes:

1. The intake valve to shut.
2. The exhaust valve to open.
3. The cylinder pressure to drop to the steam exhaust pressure—typically the 14.7 psia atmosphere for steam locomotives or river boats.
4. The piston now returns to its starting position in the cylinder, to complete its cycle.

Does this mean that the steam engine wasted 80% of the steam's energy? Unfortunately, yes! But all engines waste most of the heat in their fuel, as they convert that heat to work. That's what thermodynamics is about.

- Thermo (Heat) → Dynamics (Work)

The original text explaining thermodynamics was entitled *Heat: A Mode of Motion*, by John Tyndall, published in the 1800's and still available in paperback!

Heat and Hurricanes: Thermodynamics in Nature

Thermodynamics is the conversion of heat into velocity. We see this on the US Gulf Coast in dramatic ways. When ocean waters in the Gulf of Mexico get very warm towards the end of summer, powerful winds form (hurricanes). The warmer the Gulf waters, the more powerful the hurricane.

Heat from the Sun provides the energy for acceleration of air into wind. Heat from combustion of hydrocarbons is the energy source for the acceleration of the steam that drives turbo-generators, which turn heat into electricity.

Lessons Learned

Effective use of thermodynamics and the Mollier diagram, combined with an understanding of enthalpy (or the heat content of the motive steam), can result in both enhanced energy efficiency and an increase in turbine horsepower output. The flow control valves on the discharge of variable speed-driven pumps and compressors—and/or spill-backs—are not needed and are wasted energy. All steam turbines are originally equipped with hand (or horsepower) valves, visible on the front end of the turbine case. Every time you close such a hand valve, you will be saving about 10% of the steam turbine's steam supply.

20

Reciprocating Compression and Refrigeration

Introduction

The use of reciprocating compressors is widespread in the natural gas industry for wellhead production of gas that is largely methane. In a refinery, as gas density is reduced due to low molecular weights (i.e., high concentrations of hydrogen in hydrotreater recycle gas), the use of reciprocating compressors, rather than centrifugal compressors, is favored. The example cited in this chapter relates to optimization of the molecular weight of a circulating refrigerant vapor, to increase refrigeration capacity of a motor-driven (i.e., fixed-speed) reciprocating compressor. This example is drawn from very early in my career.

Purpose

Reciprocating compressors, in theory, have a much higher compression efficiency than centrifugal compressors. In practice, a reciprocating compressor typically has a far lower efficiency than a centrifugal compressor. The reciprocating compressor, driven by a constant-speed motor, has a variety of methods to control flow. If these various methods are not used correctly, the compressor's efficiency will be severely compromised. Also, "recips" are subject to malfunction of their inlet-outlet valves. Often, these valves malfunction and the problem is not identified by the unit engineers or the operators.

This story illustrates how to increase capacity of a refrigeration circuit, limited by the capacity of a single-stage, motor-driven reciprocating compressor. I have used the technology described (i.e., changing the volatility of a hydrocarbon refrigerant) on many process units, with great results. Using the same logic of altering the molecular weight of the refrigerant, I increased the capacity of a refinery sulfuric acid alkylation unit, when I worked in Texas City, from 18,000 BSD to 26,000 BSD, over the course of a year.

DeBottlenecking a Refrigeration Circuit

I started high school in September 1957. Mainly, I was interested in meeting girls. The first step in meeting girls is to talk to them. For three years, I waited for girls to talk to me, with no success. No one had explained that, to meet girls, I would have to talk to them first.

Now I know better.

You have to learn from experience. My career as a process engineer is a chain of such learning experiences.

Refrigeration

My problem with talking to girls was a symptom of a character flaw. For instance, as a child, one of my interests was air-conditioners. How do they work?

Our air-conditioner had several components (Fig. 20–1):

- The *compressor*, which compressed Freon vapor and pushed it around in a closed loop
- The *condenser*, which condensed the Freon from a vapor to a liquid
- The *let-down valve*, which partially flashed the Freon refrigerant from a high pressure, and warm temperature to a lower pressure and a cold temperature
- The *evaporator*, where the circulating air was cooled by the vaporizing Freon

From the evaporator, the vaporized Freon flowed into the compressor's suction and then circulated in a closed loop. I've summarized this refrigeration cycle in (Fig. 20–1).

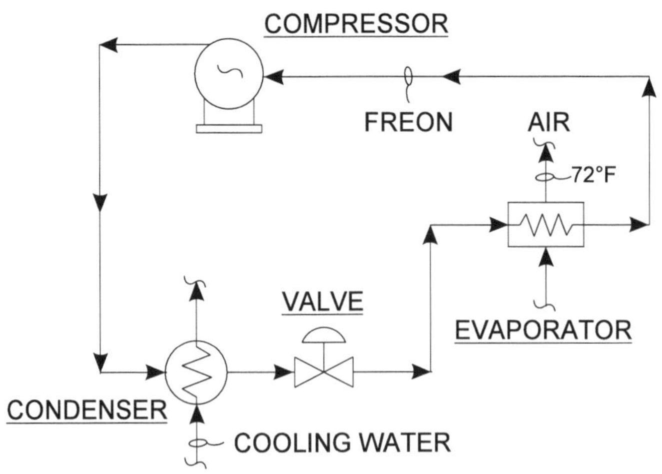

Fig. 20–1. Refrigerant circuit

In the 15 years that I worked for American Oil, I was never promoted. But why? For example, once I was assigned by Dave, our manager in Chicago, to provide the design specifications for a new centrifugal refrigeration compressor for the American Oil Refinery in El Dorado, Arkansas. "Lieberman, we only need enough detail to obtain a cost estimate for the new compressor. The existing compressor has been reported by the El Dorado refinery to be bottlenecking production of their viscous polypropylene product. American Oil will hire an experienced engineering contractor for the actual design. Your work is preliminary."

"Dave, sir, would you mind if I went to El Dorado for a few days to look the job over in the field?"

"Can't you just work up details for the estimate from files we have in Chicago? Our travel budget is limited. You can ask the unit engineer in El Dorado to mail you any missing drawings and relevant plant data."

"I was planning to drive from Chicago to El Dorado over the weekend. That will save the airfare, and I'll travel on my own time."

"Okay, Lieberman, that's fine. But just limit your expense account meals to $25 per day."

Refrigeration Limit

"Mr. Norm," Kenny, the lead operator explained, "we got only one limit on this polypropylene unit. The polymerization reaction gives off heat (i.e., an exothermic reaction). To meet our viscosity spec of 450 centistokes, we need to keep the propylene polymerization reactor at 40°F. If it gets hotter, the propylene product viscosity drops too low.

"To keep the reactor cool enough," Kenny continued, "we need to increase the refrigerant circulation rate, as we increase the propylene feed rate to the reactor.

"The refrigerant vaporizes in the polymerization reactor (Fig. 20–2). It flows as a vapor to our refrigeration compressor. The compressed refrigerant vapor is condensed back to a liquid and recirculates back to the reactor," Kenny explained. "If we run more propylene feed, to make more product, then the reactor gets hotter, because we can't circulate any more of the refrigerant. We're limited by the compressor's capacity."

"Kenny, why can't you run the compressor a bit faster?" I asked.

"No, 'cause it's a motor-driven machine. Driven by an ordinary, 3-phase, AC, fixed-speed motor that spins at 3600 rpm."

"Kenny, what's the refrigerant?"

"Mr. Norm, just ordinary LPG. Propane. We make it in the refinery and sell it to the local distributors."

"Kenny, is the motor at its max amp load? Will it trip-off on high amps if we increase the compressor load?"

"Nope, Mr. Norm. That motor is oversized for the current amp load. We are well below the motor's FLA (full limit amp load)."

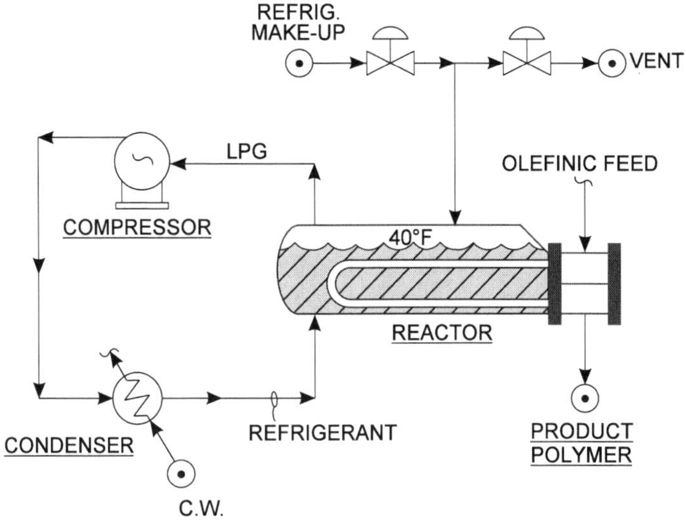

Fig. 20–2. Refrigerated reactor

My Home Air-Conditioner

I recalled that once, the air-conditioner in my home was not cooling properly. The A/C mechanic came out to my house and checked the inlet (suction) pressure to the A/C compressor. "Mr. Lieberman, you need Freon. The compressor inlet pressure is really low. You need me to add some Freon. The higher the compressor suction pressure, the more Freon will circulate (Fig. 20–1), and the more heat will be removed from the evaporator coil in your attic. And the cooler the air circulation will be in your home."

"How much will that cost?"

"It's $22 per pound of Freon-22. I'll add Freon, based on the compressor suction pressure."

So I started to think, based on my home A/C experience, how raising the El Dorado unit compressor suction pressure, to increase the propane refrigerant suction pressure, would circulate more propane. However, this would also raise the reactor temperature according to the vapor pressure curve for propane. Approximately each increase in the compressor suction pressure of 1 psi would increase the reactor temperature by about 2½°F. But I was not allowed to raise the reactor temperature, because that would degrade the product viscosity! What to do?

Altering the Refrigerant Composition

"Kenny, why do you use LPG propane as the refrigerant?"

"Because we always use propane."

"Kenny, you got any ethane in this plant?"

"Yes, Mr. Norm! We make ethane off of the FCU de-ethanizer. We sell it to the Chemicals Plastic corporation across the interstate, on the other side of town."

"How does it get there?"

"By truck," Kenny replied.

"What do you think about spiking the propane refrigerant with ethane?" I suggested. "This would:

- Increase the volatility of the circulating refrigerant. Ethane is a lot more volatile than propane.
- Increase the compressor suction pressure at a constant compressor suction temperature, because ethane has over three times the vapor pressure of propane at the same temperature.
- Increase the compressor suction vapor density, at the higher pressure. Raising the suction pressure by 30% increases vapor density by a bit less than 30%.
- Allow more refrigerant circulation because the vapor density has increased.
- Allow more propylene to react and to produce more viscous polypropylene product, because more pounds of vapor can be circulated by the compressor."

"Mr. Norm, you need to discuss this with Eric Johnson. He's the plant manager. I can't do stuff like adding ethane to the circulating LPG refrigerant on my own!"

Fig. 20–3. Norman Lieberman in El Dorado, Arkansas starting up revamped refrigeration circuit -Jan-1968

Management Approval

The next day, I explained my idea of unloading part of a truck of ethane gradually into the circulating propane refrigerant, so as to increase its ethane content from 1% to about 5%. This, I explained, would increase polypropylene production by roughly 25%, by increasing the compressor suction pressure. The plant manager was interested, but cautious.

"Lieberman, I don't think this will work! I understand your plan to vent off propane-rich refrigerant to the flare, as you add ethane from a truck. But the extra refrigerant circulation of 25% will increase the amperage load on the compressor motor also by 25%. Lieberman, it's just like my A/C at home. When I add Freon to increase cooling, that raises the compressor suction pressure. My electric meter then spins faster, because I'm pulling more amps, as Freon circulation increases. I'm afraid that you're going to trip the compressor motor on high amps! It won't work!" Mr. Johnson concluded.

"But, Sir! That motor is vastly oversized. I'm not close to the motor's FLA (full limit amp load) trip point!"

"Oh. Okay, Norm. Let's give it a try tomorrow."

Conclusion

We spiked the LPG refrigerant slowly with ethane from a truck that evening. The plant flare temporarily lit up the entire El Dorado refinery, as we gradually increased the volatility of the circulating refrigerant. By noon the next day, we had increased the viscous polypropylene production by around 25%.

I slept over at the Holiday Inn. The next day, I drove home to Chicago. When I arrived at work, I reported, rather proudly, my success to Dave. But he seemed to be angry. "Lieberman, where are the detailed specifications for the new centrifugal compressor, so that we can obtain a cost estimate? We require an estimate to obtain management approval before hiring an engineering contractor to design the new compressor for the El Dorado viscous polypropylene plant. Did you not understand the assignment?"

"But, Dave, I—"

"And where, Lieberman, are the receipts for all your meals? Many receipts are apparently missing. And the charge for use of your personal car is not allowed for weekend travel!"

I worked for American Oil from 1965 until 1980. I did not "climb the ladder of success." Then I went to work for Jack Stanley, the owner of the world's largest independent refinery in Louisiana, who was only interested in results and refinery profitability.

So often in life, we struggle against adversity, but without success. Painful as it is, sometimes it's best to cut our losses and start over. This decision proved to be the turning point of my professional career.

Low Compressor Suction Pressure Limit

At my sulfuric acid alkylation unit in Texas City, I increased alkylate (i.e., high-octane gasoline production) from 18,000 BSD to 26,000 BSD, by overcoming a refrigeration limit. The refrigeration compressor was limited by low-suction pressure. That is, the variable-speed compressor suction pressure was only 1 psig.

It's dangerous to operate a hydrocarbon refrigeration compressor suction at less than atmospheric pressure. At a slight vacuum (i.e., less than 14.7 psia), air could be drawn into the circulating refrigerant and form an explosive mixture with the hydrocarbon refrigerant in the refrigerant receiver

Always maintain a positive suction pressure on such a closed-loop refrigeration circuit. Especially if the refrigerant is a combustible hydrocarbon. In this case, the refrigerant was 90% butane and 10% propane.

The unit limit was the refrigeration circulation rate. If I circulated more refrigerant, I'd increase gasoline production. I could circulate more refrigerant by running the variable-speed, steam turbine-driven compressor faster. But then, the compressor suction pressure would drop below 1 psig—which is dangerous!

Recalling my experience in El Dorado, I slowly (over a period of a year) added propane to the circulating refrigerant. I daily increased the compressor's speed. This allowed me to circulate more refrigerant and increase alkylate production by 40%,—but without a reduction in the compressor suction pressure.

This worked because I was not limited by compressor horsepower, or the refrigerant compressor's discharge condenser's capacity, or the condenser pressure rating.

The extra 8,000 BSD of high-octane alkylate production increased gasoline supply in America by 0.13%. I was able to achieve this feat by applying the principles I had learned in El Dorado ten years previously.

Lessons Learned

Recips are best used for low-molecular (i.e., 5–15) molecular weight gas. Using a centrifugal compressor would require too many wheels and/or too many compression stages. Controlling flow through a fixed-speed, motor-driven reciprocating compressor efficiently is a complex matter, that requires the use of "Unloading Pockets".

Be alert to hot recip valve caps on the inlet and outlet. Once the inlet cap is hotter than the feed gas, the compression efficiency has been severely compromised.

21

Hydraulics

Introduction

Bernoulli's equation has always been the basis for fluid flow calculations. It relates:

- Velocity
- Momentum
- Density
- Head
- Pressure
- Frictional losses
- Pipe diameter

in a single equation. Understanding the use of Bernoulli's equation is the prerequisite to effective process engineering, operations, and design.

Purpose

The following list consists of process equipment operations that are governed by the principles of hydraulics:

- Orifice plates
- Steam vacuum ejectors
- Flow-through pipes
- Centrifugal pumps
- Piping system
- Centrifugal compressors
- Pressure drop-through holes

In My Bathtub

I always try to relate process calculations to some personal physical experience. Certainly, I rely to an extent on formulas that I learned in school or have read in books. But my calculations are going to be used to construct actual process equipment—or perhaps to make operating changes to a refinery unit after lunch today. That's why I like to relate these calculations, in a quantitative way, to some personal, physical experience.

For instance, if, as I have often done, I will be calculating the size of a draw-off nozzle from a fractionator, I like to relate my nozzle-sizing calculations to my daily bath.

Converting Potential Energy into Kinetic Energy

Have you ever taken a bath? I bathe daily after I run 5k. I love to bathe. Not because I like to be clean, but to conduct hydraulic experiments. One experiment I conduct in my bathtub relates the rate of water flow through the drain vs. the height of water in the bathtub. Often, while I'm conducting this experiment, my wife, Liz, bangs on the bathroom door. "Norman! What are you doing? You've been taking your bath for an hour—hurry up!"

"Liz! I'm conducting an experiment in hydraulics. I'm applying Bernoulli's equation to fluid flow through an orifice. Don't bother me!"

First, I'll calculate the rate of water draining out of the tub. My tub is 4 ft. x 3 ft = 12 ft^2 If the water drops by 2 inches, that's ⅙ of a foot. So, the volume outflow is 12 ft^2 x ⅙ ft = 2 ft^3

It takes a minute for the water level to drop by this 2 inches. So, the rate of flow through the bathtub drain is 2 ft^3 per minute.

My bathtub drain is 1 inch in diameter. The area of a 1-inch hole is 0.005 square feet. That calculation is:

- $(½)^2$ • (3.14) ÷ 144 = 0.005 sq. feet.
- The "3.14" factor is pi.

That is, the area of a hole equals half the diameter squared x pi (square inches), divided by 144 square inches, per square foot.

Then, if I divide the rate of volume flow (2.0 ft^3/minute) by the area of the hole (0.005 ft^2), the water velocity through the hole is (2.0) ÷ (0.005) = 400 ft. per minute, or:

- (400) ÷ 60 = 6.6 feet per second = V (i.e., 60 seconds per minute).

According to Bernoulli's equation,

- Delta H = (K) • $(V)^2$

Delta H is the depth of water in my bathtub, which was 16 inches. So:

- $16 = K \cdot (6.6)^2$
- $16 \div (6.6)^2 = K = 0.36.$

"K" is called the "orifice coefficient." When water drains through a hole, its velocity increases, or the water accelerates from being stagnant, to 6.6 feet per second. According to Dan Bernoulli, it takes energy to accelerate the water.

When you accelerate your car, that energy comes from your car's engine. When the water accelerates through the drain in your bathtub, that also takes energy. This energy comes from converting the head of water (potential energy) to velocity (kinetic energy).

Frictional Losses

The problem with my bathroom calculation is that the theoretical "K" factor in Bernoulli's equation is 0.18—not my observed 0.36. Why the difference?

The difference is due to friction and turbulence, as the water flows through my bathtub drain. The orifice coefficient is typical. I've used this 0.36 value in my design for flow-through nozzles for the last fifty-eight years. However, on occasion, I like to verify the nozzle orifice coefficient by a field experiment in my bathtub. Don't take my word for this; try it yourself.

Variations in Orifice Coefficients

Many drains or vessel draw-off nozzles are equipped with vortex breakers to retard vortexing. This increases the orifice coefficient, but I do not know by how much. It depends on the design of the individual vortex breaker.

For a sharp-edged, thin orifice plate, a typical coefficient might be 0.6. The thicker the plate, and the more rounded the opening, the smaller the coefficient.

In process equipment, the orifice coefficients I have observed for refinery nozzles are closer to 0.20–0.25 than my bathroom experimental results of K = 0.36.

When I worked for American Oil, our company policy was to use a nozzle exit loss coefficient of 0.34 (with or without a vortex breaker).

Centrifugal Pump Impeller and Volute

In a centrifugal pump case, we convert velocity to feet of head. The liquid flows into the center (eye) of the impeller. As the liquid flows through the impeller, its velocity increases from perhaps 5 ft/second to perhaps 100 ft/second. As the liquid leaves the pump discharge nozzle, which is cone-shaped (with the wider side at the end of the cone, which flanges up to the discharge line), the kinetic energy provided to the liquid from the impeller is converted into "feet of head."

Feet of head, multiplied by the density (or S.G.) of the liquid, is proportional to the pump's differential pressure:

- $\text{Delta P (psi)} = \dfrac{\text{(Feet of Head) (S.G.)}}{2.31}$

The reader should draw three important conclusions from the above function of the volute in a centrifugal pump:

- The pressure inside the pump's case is not the pump discharge pressure, but the pump's suction pressure. Most operators do not know this.
- When the flow through a pump is reduced by excessively throttling back on the pump discharge valve, the pressure inside the case will rise to the observed discharge pressure. This is bad for the pump's mechanical seal integrity. Be careful to avoid running a centrifugal pump too far back on its performance curve.
- The pump's discharge pressure is linearly proportional (for non-viscous flow) to the specific gravity of the liquid being pumped.

Pressure Recovery from Control Valves

I once measured a pressure drop through a mostly closed 2-inch FRC (flow control valve) of 40 psi, in a diesel oil product line. The upstream pressure of the FRC was 140 psig, and the downstream pressure was 100 psig.

The diesel oil flowed into a heat exchanger, about 100 feet downstream, at the same elevation as the control valve, through a 4-inch line. Much to my surprise, the inlet pressure to the heat exchanger was 120 psig—that is, 20 psi higher than the outlet pressure of the 2-inch flow control valve. What had happened?

Pressure Drop Surveys

The velocity of the diesel oil flowing through the 2-inch opening of the flow control valve had dropped from perhaps 10 feet per second to about 2 feet per second in the 4-inch run-down line. The reduction of velocity of 8 feet per second had accounted for my observed pressure increase.

I call this "pressure recovery." It's a problem that I keep in mind when I run pressure drop surveys in the field, downstream of all control valves. It's a rather common error when running a pressure drop survey.

Lessons Learned

This chapter explores the relationship among these types of subjects:

- Momentum
- Feet of head
- Head pressure
- Pump NPSH available
- Frictional losses in piping
- Fractionation tray efficiency
- Measurement of flows
- Centrifugal pump operation

The chapter is best viewed as an introduction to the largest branch of applied engineering, discussed in greater detail in the following two chapters.

22

Pressure and Pressure Drop in Pipes

Introduction

This is a follow-up chapter to Chapter 21 "Hydraulics." Piping pressure loss is a function of three variables: friction, acceleration, and head. As a fluid accelerates through a pipe, its pressure is converted to kinetic energy or momentum. As a fluid gains elevation in a pipe, pressure is also lost. Depending on surface roughness, frictional losses in piping may be very large, but never negligible.

Purpose

Use of the standard charts that are widely available to calculate pressure drops in pipes may be subject to a wide range of errors. These charts are based on piping losses measured in piping systems in England many years ago. Therefore, there is an assumption about the piping interior surface roughness that may—or may not—apply to your current application. Better to measure the piping losses in an existing application, similar to your projected situation, and ratio the observed pressure drop to the new application.

The pressure at my home in New Orleans is 0 psig or 14.7 psia. I live at the bottom of an ocean of air. The column of air extends up tens of thousands of feet. The density of this air drops by half at an elevation of 25,000 ft. above my head and by 90% at 50,000 ft. However, at sea level, if you measure the weight of this column of air on a 1-inch square surface, that weight will be 14.7 pounds per square inch. The "A" in psia indicates that no adjustments have been made for these variables:

- Elevation
- Atmospheric temperature
- Weather conditions

The "A" stands for "absolute." Technical calculations are always based on the absolute pressure. "psig" measurements are called gauge measurements and are not used for technical calculations involving gas or vapor flows.

Water Pressure

Water is much denser than air—about 800 times denser. Because water is so dense, it creates a lot more head pressure on a square inch of surface area. A water height of only 28 inches will create a pressure of 1.0 psi. (The water pressure at the deepest portion of the Pacific Ocean, in the Mariana Trench, is about 11,000 psi. Sea water is 3½% more dense than fresh water, due to its salt content.) A water height of 34 feet will create a pressure equal to the many tens of thousands of feet of air in our atmosphere.

The pressure of city water in your home is around 30 psig—or 30 pounds per square inch—above the pressure of the air in your house. What maintains this water pressure?

Observe the water tower near your home. Count the rungs on the water tower's ladder. If your home water pressure is 30 psig, the ladder will have 70 rungs (each ladder rung is one foot apart). This elevation will produce a head pressure of 30 psig.

Each 28 inches of water will cause a pressure of 1 psi. This assumes that the water is not mixed with air or gas. Bubbling air through a column of water reduces the density of the water. This results in a reduction of the head pressure that a given height of bubbling water will develop.

Pressure Drop

Water flowing upwards in a pipe will lose head pressure as it gains elevation. Each height increase of the flowing water of 28 inches causes a reduction in the head pressure of 1 psi.

Water flowing through a pipe, at a constant elevation, will also lose pressure due to friction. A shiny, stainless-steel pipe will not create much frictional pressure loss. An old, corroded, carbon-steel pipe will create a lot of frictional loss, especially at a high velocity.

Reducing the inside diameter of a pipe, without changing the rate of the volume of flowing water, will have a dramatic effect on the pressure drop of the water flowing through the pipe. For example, I have 10 GPM of water flowing through a 4-inch pipe. Let's assume that the pressure drop is 10 psi.

If I now switch to a 2-inch pipe, the pressure drop of water in the pipe will increase from 10 psi to 320 psi. This is the rule:

- Pressure drop varies with the pipe I.D., raised to the fifth exponent (or power).
- That is $(4''/2'')^5 \times 10$ psi = 320 psi.

This does *not* include the energy needed to accelerate the water. It does include the extra friction, due to the increase in the ratio of the water flow to the pipe diameter—and may be on the conservative side.[IV]

Pressure to Heat

The pressure lost by the water flowing through a pipe is converted to another type of energy. That's HEAT. Water flowing, at a high velocity, through a long, rough pipe will noticeably get hotter. A relatively small increase in temperature of flowing water indicates a relatively large loss in flowing pressure. This will occur at a higher pipe velocity. A high pipe velocity is 15–18 feet per second. A low velocity is 2–3 feet per second. Velocities above 18 feet per second are almost certainly erosive.

Another factor that contributes to high pressure in pipes is viscosity. Water has a low viscosity, as do most other liquids such as gasoline, propane, and kerosene, even at low temperatures. Other liquids, especially when cool, have high viscosities. For instance, baby oil, lube oil, and tar can have super high viscosities, unless hot.

Maple syrup, when cold, has a really high viscosity, but a low viscosity when hot. Changing viscosity has a very, very large effect on pressure drop of a liquid flowing through a pipe.

The density of a liquid has a minor effect on pressure drop. Water is 30% denser than gasoline. At a constant velocity, changing from water flow to gasoline flow in a pipe will only reduce the fluid pressure drop by 30%.

Valves and Elbows

Pipes with elbows and open valves contribute to increased pressure drop. Typically, a pipeline that has a number of elbows and open valves will have double the pressure drop of a pipeline of the same length that is straight and free of valves. Sharp turns in a pipe may double pressure loss. Wide-open globe valves have a lower pressure drop than gate valves. Changes in a pipe's diameter also create more pressure drop.[IV]

On the other hand, changes in the flowing water elevation, if the terminal elevation is the same as the starting elevation, will not cause an increase in pressure drop of the flowing water. At the high point of a piping run, the pressure may be far lower than the terminal pressure.

Converting Pressure Back to Height

The water pressure in your home is derived from the elevation of the water in the local water tower. You can convert this pressure back into head by attaching a long water hose to a faucet and then climbing a ladder equal in height to the water

IV. King, H.W. and Brater, E.F. *Handbook of Hydraulics*, 5th edition (New York: McGraw-Hill Book Company, 1963).

tower. This is not exactly true when the water use in your neighborhood is high, as the frictional losses in the area's distribution piping will diminish the conversion of pressure in psia back to head, in feet.

Converting Water Pressure to Velocity

The nozzle on the end of a garden hose is a common example of the conversion of water pressure to velocity. Decreasing the water pressure through a nozzle by a factor of four will increase the nozzle's exit water velocity by a factor of two:

- Pressure drop of water is proportional to the increase in water velocity squared:
    - DP proportional $(\text{Velocity})^2$

This relationship neglects the effects of friction in the hose and the nozzle.

The conversion of water's elevation to pressure, and then to velocity, is how we drive water turbines to make electricity. It's the velocity of the water striking the turbine's wheels that causes the turbine rotor to spin, which then generates electric power.

The Hoover Dam is an illustration of this hydraulic principle in action. Squirting water against the spokes of your bicycle's front wheel, resting on the seat and handlebar in your driveway, is a more convenient way to illustrate the principle of conversion of water pressure to rotational energy, which can then spin a turbo-generator.

Much of what we do in the process industry is converting pressure into velocity, or velocity into head, or head into pressure. At the Hoover Dam, water head is converted first into pressure, then into velocity, and finally into electric power.

Frictional Losses in Piping

A liquid flowing through a pipe loses pressure for several reasons:

- The length of the pipe
- The velocity in the pipe
- The roughness of the inside of the pipe
- An increase in elevation of the pipe
- An increase in the velocity of the liquid flow because the pipe size decreased
- Changes in direction of flow due to elbows
- High viscosity
- High density

A typical liquid velocity in a pipe is 4–12 ft/sec. Velocities of 16–18 ft/sec. cause erosion and are not recommended. A really high viscosity is 60–100 centipoise. This will greatly increase pipeline pressure drop. A typical pipeline pressure drop is 1 to 5 psi per 100 feet.

Using Pressure Drop Correlations

The standard method to calculate pressure drops in long lengths of pipes is prone to error. These correlations all depend on the selection of friction factors.[IV] The friction factors are calculated based on the roughness inside the pipe, which is variable and rather impossible to ascertain.

My method is quite simple and more likely to produce a correct result. That is, select a length of piping currently operating in a similar service. Measure the pressure drop in the field over a few hundred feet of straight piping. Then, back-calculate the friction factor required to force the measured pressure drop to agree with the pressure drop observed in the field.

I always attempt to extrapolate from measured variables, which I then use to adjust the standard process correlations. My technique is much harder to employ than using standard industry correlations or computer-generated calculations. The trick to making my approach to technology work is to make all field measurements yourself.

Lessons Learned

Piping pressure drop varies not only with "equivalent feet," but with the following:

- Valves and fitting in lines
- Frictional losses depending on interior surface roughness
- Changes in velocity
- Fluid viscosity
- Fluid density
- Changes in elevation

Flowing pressures often increase in piping systems, if the flowing fluid slows down, as the flow enters a pipe of increased diameter.

23

Vacuum System and Vacuum Surface Condenser Problems

Introduction

Troubleshooting vacuum systems is my main area of activity. Most vacuum systems are designed and manufactured by the Graham Corporation in Batavia, New York. Vacuum systems are very complex and consist of the following components:

- Vacuum steam ejectors
- Surface condensers
- Seal legs
- Hot well (or condensate drum)

Also, they include liquid seal ring pumps, which is an alternate to a vacuum steam ejector or jet. I have written a separate book on this subject: *Troubleshooting Vacuum Systems*. Wiley, 2012

Purpose

The steam jet is actually a two-stage compressor. The off-gas from the vacuum tower or the air from the steam turbine vacuum surface condenser is drawn into the inlet of the steam jet mixing chamber. The steam jet has no moving parts. The energy to compress the off-gas from the vacuum tower is derived, not from the pressure of the motive steam, but from the enthalpy—or heat content—of the motive steam.

"Mr. Norm," Dave asked, "could you help me out with #2 vacuum tower? The flash zone's running at 55 mmHg. We got 20%, 1050°F minus for our D-1160 ASTM distillation analysis of the vacuum tower bottoms residual oil. Used to be around 35 to 40 mmHg and 10%–12%, 1050°F and lighter. We're losing 3,000 BSD of heavy gas oil to vacuum bottoms resid. At $20 a barrel downgrade to coker feed, we're losing $60,000/day."

"When did this start getting so bad?" I asked.

"It's been gradual, Mr. Norm, over the past year."

"Dave, vacuum systems are complicated. They're hard to understand.

The following are some of the problems that cause a progressive loss of vacuum:

Motive Steam Nozzle Malfunctions

"Silicate hardness deposits accumulate at the nozzle throat. This restricts the steam flow and leads to a loss of the 'sonic boost.' Or the jet is forced out of 'critical flow,' which means the same thing."

"What's the 'sonic boost,' Mr. Norm? Is that good or bad?" Dave asked.

"It's good. You see, Dave, the motive steam flows into the steam jet at one or two hundred feet per second. It blasts out of the steam nozzle at 10,000 to 12,000 feet per second (Fig. 23–1). That's ten times sonic velocity. But, at some point in the diverging part of the steam ejector, the steam slows to subsonic. The vapors catch up to each other. Like a sudden traffic jam. The kinetic energy of the vapors is converted to pressure. That's the sonic boost. I've seen the resulting compression of the gas be as much as 12 to 1. But 7 to 1 is more typical, as shown in (Fig. 23–1)."

"So, Mr. Lieberman, the ejector is like a compressor. But without moving parts? What then is compressing the flow?"

"The velocity of the steam, Dave. As it slows, its kinetic energy is suddenly converted to pressure."

"But where, Mr. Norm, does the kinetic energy of the steam come from? From the steam's pressure?"

"No, Dave. It comes from the heat of the steam.[V] Look at (Fig. 23–1). The steam cools from 350°F down to 100°F as it flows through the steam nozzle. That's called an 'isentropic expansion.'"

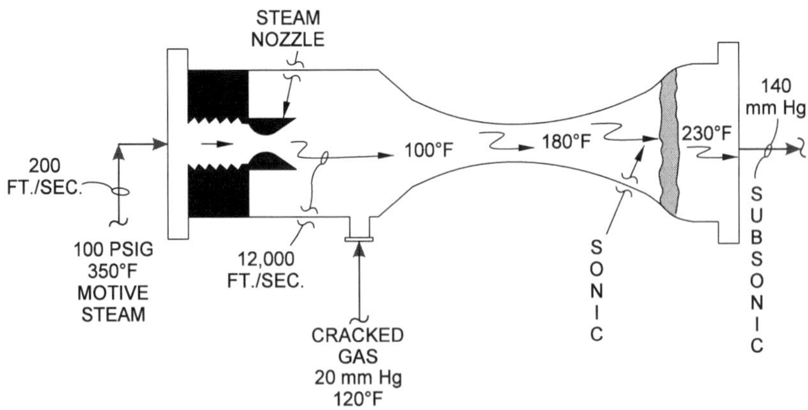

Fig. 23–1. Steam ejector developing sonic boost

V. Power, Robert. *Steam Jet Ejectors for the Process Industries* (McGraw Hill, 1994).

"I think we learned about that in thermodynamics at LSU. But sometimes that jet starts to make a surging, sick kind of sound. And our vacuum pressure gets real bad all of a sudden. The operators say, 'the jet is breaking.' The vacuum tower top pressure can jump from 20 mmHg up to 40 mm. Is that caused by silicates plugging up the steam nozzle's throat?"

"No, Dave. That 'surging' or 'hunting' sound means you have lost the sonic boost, and the ejector is no longer in its critical mode of operation. The vendors call this the 'break-and-pick-up mode of operation.' Most often, that's caused by the jet discharge pressure getting above its 'maximum allowable discharge' pressure. That's shown on the ejector's performance curve. Maybe the problem is an excessively high ejector discharge pressure, because the downstream condenser is fouled. Or maybe it's the—"

"But, Norm, so it's not just the steam nozzle getting fouled with hardness deposits from the motive steam?"

"Look, Dave. Let me explain. It's kind of complicated."

Source of Silicates in Steam

"I guess, Norm, the silicates in the steam are caused by poor-quality boiler feed water?"

"No, not usually. It's more likely that the boiler is suffering from poor level control. Water, full of silicates, carries over into the steam."

"Well, I can't help that. What can I do if that's my problem?"

"Dave, often there's a clean-out plug in the back of the ejector. You can shut down the jet, remove the plug, and rod out these deposits."

"That's pretty simple! What else?"

"The second-most common cause of bad ejector performance is the steam nozzles get loose. The nozzles are made out of 316 stainless steel, but are screwed into a carbon-steel ejector body. Since our steam is wet, galvanic corrosion results. The carbon steel threads get corroded out, and then the steam blows out around the nozzle."

"Guess that means we should buy a new ejector, Mr. Norm?"

"Certainly not! Just wrap the nozzle threads with Teflon tape, or a Teflon pipe thread-type paste, and screw the steam nozzle back in real tight."

"Maybe that should have been done in the factory, I reckon, Norm?"

"The next most common problem, Dave, is wet steam. A little bit of moisture is no big deal. But if the steam is real wet, it can make a big difference.[VI] If the steam blows out of the bleeder white, at the tip of the bleeder, it's wet. Dry steam is invisible.

VI. Lieberman, Norman P. *Troubleshooting Vacuum Systems* (Scrivener Publications, John Wiley & Sons, 2012).

"Here's what I do: Make sure the 'de-superheating station' is on auto. Set it for 10°F to not more than 20°F above saturation temperature. Not like the Pemex folks did in Monterrey. They set the de-superheater for 5°C below saturation. I blocked in the condensate flow, and the vacuum tower top pressure improved from 35 mmHg to about 25 mmHg. Actually, 10°F of superheat is no big deal. Best to just ignore superheat unless it's 20°F plus, Dave."

"And wet steam, Norm. Is that really so bad?"

"Well, yes. I don't know how to quantify the effect. If the steam looks a little white, it probably is okay. If it blows out of the bleeder dripping wet, that could be a big problem. Once, I wrapped 20 feet of 2-inch steam line with rags at the Hunt Refinery in Tuscaloosa, Alabama, and gained 3 or 4 mmHg tower top pressure. As Hunt was borderline on their paving asphalt hardness spec., this was a significant improvement."

Motive Steam Pressure

"How about steam pressure. What's best—higher or lower?"

"Good question, David. You would think higher-pressure steam would help. But my experience teaches that typically reducing steam pressure improves vacuum, when the maximum allowable discharge pressure is exceeded. That means your discharge condenser is overloaded. If you reduce the supply steam pressure by 10%, then the condenser load will drop by about 10% and maybe allow the jet discharge pressure to drop below its maximum allowable discharge pressure. I did this once at the Hunt Refinery and upgraded 9,000 BSD of vacuum resid from delayed coker feed to paving grade asphalt. Worth like $30/barrel to Hunt."

"Wow, Mr. Norm. $270,000 a day! They paid you a really big fee, I bet!"

"No. It don't work that way. Then there was the time I was working in Delaware for Texaco where I improved vacuum from 18 mmHg to about 15 mmHg by increasing the motive steam pressure from 140 psig to 160 psig. The higher-pressure steam stopped the jet from surging. They were controlling to the design motive steam pressure of 160 psig. But the pressure control valve (PRC) was at grade, and the steam jet was about 100 feet high and several hundred equivalent feet of piping away."

"You must have figured that out right off, Mr. Norm."

"Nope! I sat on the sixth landing all alone in the dark and cold, with a nasty wet wind coming off the Delaware River, until midnight, before I reset that PRC up to 180 psi. It ain't being smart that counts, Dave, with vacuum systems; it's being persistent."

Seal Drum Problems

"Dave, there's a whole class of problems relating to the seal drum and the barometric drain legs (Fig. 23–2). These problems cause liquid backup in the surface condensers. The liquid covers the tubes and raises the condenser pressure above the maximum allowable discharge pressure (MADP, shown on the vendor's performance curve data sheet).

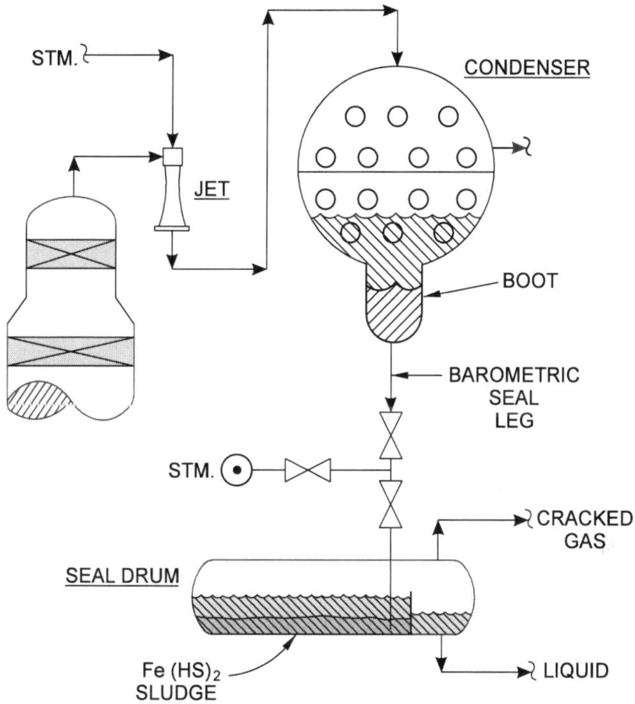

Fig. 23–2. Condensate back-up raises condenser pressure.

"One common problem is an air leak in the barometric leg. The air gets sucked into the leg and interferes with condensate drainage from the condenser boot."

"How do you find leaks like that, Mr. Norm?"

"Cold spots! Air drawn through a hole from the atmosphere into a vacuum cools off as it expands."

"I know! That's a Joule-Thompson expansion. I learned 'bout that at LSU. You got to shut down to fix them leaks, I guess?"

"No. Duct tape, Dave. Works especially good for leaking flanges. I fixed this type of problem at the Mobil Refinery in Coryton, UK. I wrapped a flange on a steam turbine exhaust surface condenser with tape. Debottlenecked the FCU air blower. The big problem with the seal drum is sludge. Mainly iron sulfides ($Fe(HS)_2$). It covers up the bottom of the seal leg from the first stage condenser. Then it can't drain properly."

"This must take a shutdown to clear out the sludge."

"Nope. Here's a trick I learned from the operators at the Sinclair Plant in Wyoming (Fig. 23–2). Just blow out the seal leg with steam to clear away the sludge and restore condensate drainage. You can know when you have to do this steam-out by looking at the temperature difference between the condenser vapor outlet and the boot. When the boot is 10°F–20°F colder than the vapor outlet nozzle, it's time to blow out the seal leg. The guys at Sinclair did this once a week. The whole problem can be avoided by cutting off the seal leg so that it's about 24 inches above the bottom of the drum, but still a little below the overflow baffle—and by cleaning the seal drum during every turnaround.

"At the Shreveport Calumet Specialty plant, the seal drum had a level control problem. The damn drum was showing 60%, but was really full-up. I lowered the level set-point, and the condensate drained out of the condenser. The vacuum tower pressure dropped like a rock. The level-trol control had been calibrated for water (1.0 S.G), but the drum was mostly full of light naphtha (0.6 S.G)."

Air Leaks

"Mr. Lieberman, I think air leaks anywhere could overload the jets. How can I decide if I got a big air leak?"

"Get a bicycle tire inner tube and squeeze out all the air. Then reinflate it with gas from the top tap of the seal drum level glass. Be careful! That's 10%–40% H_2S in the seal drum. A concentration of 1% will kill you! If the gas chromatograph result shows less than 10% N_2, forget about air leaks. If it's 20% N_2 or more, you got a problem. If you got 10% O_2 with 40% N_2[VII], the air leak is likely in the vacuum system itself. If there is 10% CO_2 and no O_2, the air leak is likely in the vacuum heater transfer line. This is very dangerous.

"Dave, the other thing that can overload the vacuum system is too much cracked gas. Most of the cracked gas is formed in the vacuum heater tubes. To suppress thermal cracking, increase the heater pass coil steam by 1 pound per barrel.

"Sometimes much of the cracked gas is formed in the vacuum tower bottom boot. Either lower the level by 20% or reduce the boot temperature by 20°F, with more boot quench, and see if this helps.

"But maybe it ain't cracked gas that's the problem, Dave. It may be virgin C_3s and C_4s that aren't getting stripped out of vacuum tower feed. Try raising the crude tower bottom stripping steam flow by 20%–30% and see if that helps. I like to use about 0.15 pounds of steam per gallon of vacuum tower feed.

VII. Putman, R.E. *Steam Surface Condensers* (New York: ASME Press, 2001).

Boot Quench Exchanger Leak

"I visited the Repsol Refinery in Spain last year. I found that the boot quench exchanger (Fig. 23–3) was leaking boiler feed water into the boot quench return line and overloading the first-stage ejector. It was then bypassed, and the tower top pressure dropped from 45–50 mmHg down to 30–35 mmHg.

"And then there's a whole bunch of problems with the surface condenser internal vapor baffles, which if not adjusted correctly when the tube bundle is installed, can—"

"Mr. Lieberman, I can't remember all this. Can't we ask an expert to troubleshoot the problem for us?"

"Dave, there's only one thing you have to understand when troubleshooting vacuum systems. You got to take your tools and your determination out to the unit. That's it. It's all a matter of time and determination."

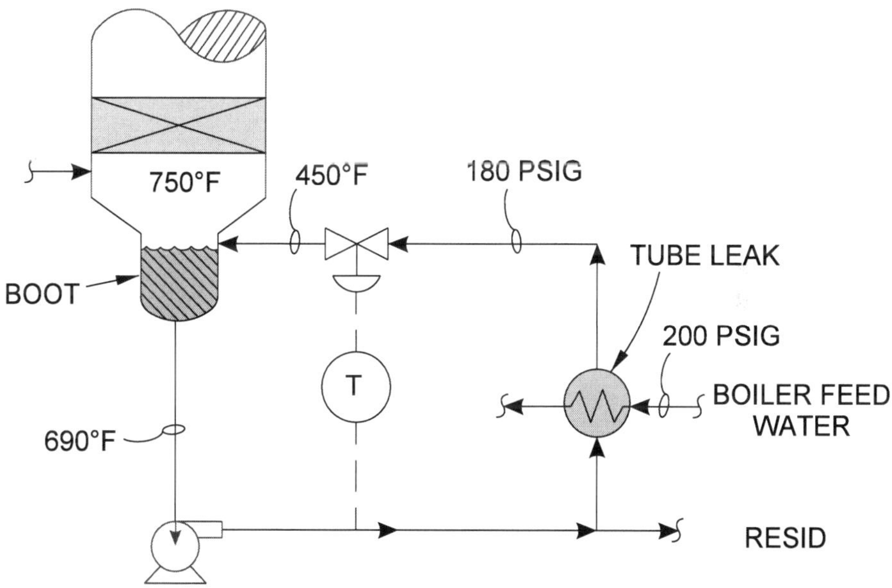

Fig. 23–3. Tube leak results in loss of vacuum due to steam overloading vacuum system.

Velocity vs. Pressure in Steam Ejector

Fig. 23–4 shows the velocity and pressure profiles in a converging-diverging steam ejector. Note that a reduction in the steam velocity is mirrored by an increase in the flowing pressure. Over half of the ejector pressure rise (the sonic boost) occurs just downstream of the narrowest portion of the jet, where the velocity of the vapor flow drops from sonic to subsonic. If back-pressure from the condenser pushes the transition from sonic to subsonic velocity back too far, the sonic boost is completely and suddenly lost.

The suction pressure to the jet (20 mmHg in Fig. 23–4) may then suddenly increase from 20 mm to 40 mmHg, and the jet will run somewhat more quietly. The jet is then said to have "broken," and may start to surge. The jet discharge temperature will also drop by about 40°F, due to the reduced compression ratio.

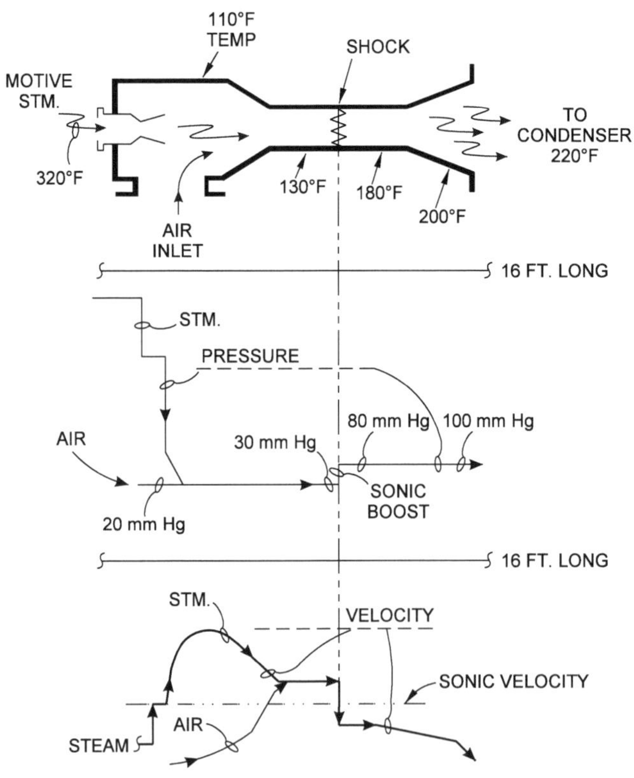

Fig. 23–4. Velocity vs. pressure profile in steam ejector. Temperatures observed on outside surface.

Common causes of poor ejector condenser performance are:

- Silicates in steam nozzle throat
- Loose nozzle
- High discharge pressure
- Wet motive steam
- Sludge in seal drum
- Excessive motive steam pressure
- Too much cracked gas
- Back-flow through parallel jets
- Air leaks in upstream vessel
- High level in seal drum
- Steam nozzle erosion
- Wax in barometric drain line
- Low-motive steam pressure
- Air leak in barometric drain line
- Steam nozzle installation error
- Excessive wear in diffuser
- Excessive steam superheat
- Excessive condensate flow

Vacuum Tower Surface Condenser

About half of the problems I encounter with vacuum towers are caused by the poor performance of the surface condensers located on the discharge or inlet side of the ejectors.

A common problem encountered in vacuum towers is excessive back pressure at the ejector discharge, due to low heat transfer coefficients in the downstream surface condensers. These are the main problems noted with these surface condensers:

- Shell-side condensate back-up
- Tube-side water fouling
- Vapor binding on the shell side

While most of the vapor binding in refinery vacuum towers is due to cracked gas rather than air leaks, the literature describing this problem is based on air leaks in steam turbine exhaust vacuum surface condensers.[VI] Whether one is dealing with hydrogen, methane, ethane and H_2S, or nitrogen and oxygen, the problem and the solution are much the same.

Surface Condenser Air Baffle

A typical design "U" value (i.e., heat transfer coefficient) for a refinery vacuum tower surface condenser may be 180 BTU/H/°F/ft^2 (service) and 300 BTU/H/°F/ft^2 (clean). In practice, even for clean surface condensers, observed "U" values are 40 to 80. Much of the problem appears to be due to leakage of steam around the leaf seals or seal strips (Fig. 23–5). This allows the condenser inlet steam to bypass most of the tubes and overload the downstream ejector.[VIII] Here is the best way to retard this leakage:

- Seal strips should extend ¼" to ½" beyond the I.D. of the shell before the tube bundle is reinserted into the shell, so as to slightly crush the strips against the shell I.D.
- Seal strip material should be 316 stainless steel, of a special flexible grade, and never copper-based alloys, which are likely to corrode in refinery service.
- Three pairs minimum should be used.

In addition, the design of the tube support baffles should allow the vapor trapped just below the air baffle, shown in (Fig. 23–5), to flow without hindrance to the ejector. Retrofitting an existing condenser shell can ensure that this objective will be accomplished.

Shell side DP should be less than 5 – 8 mm Hg.

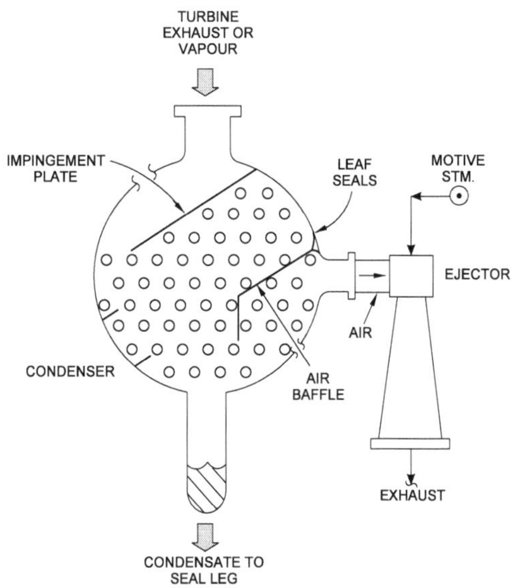

Fig. 23–5. Vapour binding increases surface condenser operating pressure and reduces the turbine horsepower.

Cooling Water Inlet Location

The performance of vacuum tower pre-condensers was greatly enhanced in the Coastal Refinery in Aruba by reversing the cooling water flow. The general

VIII. Lieberman, N.P. "Troubleshooting Steam Ejectors," PTQ, Q4, 2014, pgs. 125-129.

rule for condensers is to have the cold water inlet on the bottom of the channel head and the warm water outlet on the top of the channel head. For vacuum system condensers, the correct practice is to have the cool water inlet as close as possible to the vapor outlet nozzle that is flowing to the downstream ejector. This will minimize the vapor outlet temperature to the ejector. The lower the condenser's vapor outlet temperature, the less the water vapor content of the ejector inlet vapor. This unloads the ejector, which will then develop a greater vacuum.

There should be no concern about an excessively high liquid outlet temperature. This temperature does not control the vapor outlet flow. Indeed, a depressed liquid drain temperature is a sign of condensate backup. Submerging any condenser tubes in condensate is a definite and common way to degrade the condenser's performance. Typical causes of condensate backup include the following:

- Leaking seal legs
- Sludge in the seal drum
- High seal drum liquid level
- Excessive seal drum pressure
- Wax in the seal legs

If the liquid level rises to cover the lower edge of the air baffle (Fig. 23–5), there will be a sharp break in the vacuum. This was experienced several years ago in a plant in northern Louisiana. The cause was the lack of sufficient capacity in the seal drum's condensate pump. Thus, a high seal drum liquid level backed up the seal legs into the condenser.

Common causes of poor vacuum condenser performance are:

- Air baffle seal strips not wide enough
- Air baffle seal strips constructed from brass or copper – not 316 S.S.
- Impingement plate too big
- Condensate backup (seal leg problems)
- High cooling-water temperature
- Fouling
- Low cooling-water flow
- Use of triangular tube pitch
- Use of ¾-inch tubes on 15/16" triangular pitch
- Tube leaks
- Use of "fixed tube sheet" design tube bundle
- Channel Head pass partition baffle leaking
- Excessive elevation of condenser
- Improper bundle installation – seal strips crushed
- Air evolution from cooling water, at outlet

Third-Stage Surface Condenser

In most vacuum towers, the third-stage surface condenser has three baffles between the liquid drain and the vapor outlet:

- An underflow baffle
- An overflow baffle
- A second underflow baffle

This configuration tends to trap sludge and restrict the pathway to the vapor outlet nozzle. If you measure a high DP (i.e., over 100 mmHg) across this final surface condenser, the build-up of sludge is certain to be your problem.

I consider this configuration to be a design error. The third-stage condenser should be designed similar to the first- and second-stage condensers, with a horizontal air baffle and seal strips.

Safety Note

A sudden increase in the CO_2 content of the seal drum off-gas is an indication of an air leak in the transfer line between the heater and the vacuum tower. This line is normally under a slight vacuum. But if the vacuum is lost, then the leak will go positive. Vacuum tower feed will then blow out of the defective transfer line, and a large fire may result.

This has happened to me three times. On one occasion, the vacuum unit was destroyed. The problem was highly erosive velocity in the transfer line. In general, 347 (L) stainless steel is the material of choice for this service, not 410 chrome steel.

A sample every few weeks from the seal drum is recommended, to monitor CO_2 levels.

Lessons Learned

The performance of a steam jet can only be understood on the basis of thermodynamics. It is an illustration of the conversion of heat (enthalpy) into work or, in this case, compression of low-pressure gas or air. When you actually understand how steam ejectors work, you will really understand thermodynamics. To me, this is the most fascinating aspect of the application of our technical education, as it illustrates an isentropic expansion of the ejector's motive steam

24

Corrosion

Introduction

The common origins of corrosion in an aqueous phase in refinery and process plant applications are HCl, CO_2, and SO_3. I have not included H_2S in this list because H_2S, when not interacting with any of the above three molecules, is not particularly corrosive. Also, the above three acids are not corrosive either, unless an aqueous phase (water) is present.

One of the main products of corrosion has never been seen and can only be observed indirectly. That is ionic hydrogen—or a proton. It is these protons that are the culprit of the most dangerous aspect of corrosion. That being, "hydrogen-assisted stress corrosion cracking."

Purpose

This is the corrosion mechanism that is most familiar to refinery process engineers:

- $HCl + Fe = H^+ + FeCl^-$
- $FeCl_2 + H_2S = Fe(HS)_2 + HCl$

The end products of the above, being an iron sulfide black sludge and a hydrogen ion, cause hydrogen-assisted stress corrosion cracking.

My major problem as an operating supervisor in Texas City was corrosion. Most of the corrosion products I encountered were a black, slippery, finely divided powder. The black powder would, upon drying, autoignite. It burned with a pale blue flame, not visible in sunlight, which emitted a white vapor that made me choke. That is, sulfur dioxide (SO_2).

Analysis indicated this black powder was iron sulfide (FeHS) or pyrophoric iron. A rather surprising observation, as hydrogen sulfide, which is a weak acid, is not particularly corrosive at moderate temperatures. Why then, I wondered, was there all this corrosion to piping, vessels, and heat exchangers, operating at less than 200°F, due to hydrogen sulfide (H_2S)? The source of my corrosion problem was not so much H_2S as the real agent of corrosion—rather, it was hydrochloric acid.

Origin of HCl in Refineries

I have always considered hydrochloric acid to be a rather evil molecule. In process units, it originates in two benign ways: from sea water and to promote improved octane in gasoline.

Sea water consists of 97% water and 3% salts. The salts are mainly these three:

- 90% NaCl
- 4% $CaCl_2$
- 6% $MgCl_2$

The NaCl is pretty much removed in the crude unit desalter, as is much of the $CaCl_2$. The $MgCl_2$ (magnesium chloride) is less soluble in the desalter wash water, and hence 30–40% escapes from the desalter.

At 500°F or hotter, the $MgCl_2$ will dissociate into

- $MgCl_2 + H_2O = Mg(OH)_2 + HCl$

The second origin of HCl encountered in a refinery is manmade. We inject chlorides into the hydrogen recycle gas on naphtha reformers, to enhance gasoline octane. The chlorides react with the recycle hydrogen to produce

- $H^+ + CL^- = HCl$

Both of the above reactions produce small amounts of hydrochloric acid. HCl is completely non-corrosive, as long as it is dry. However, should the HCl—which is a very strong acid—become wet, it becomes terribly corrosive to carbon steel piping, vessels, and heat exchanger tubes. The result of this corrosive reaction is

- $2HCl + Fe = Fe(Cl)_2 + H_2$

The hydrogen is produced initially as a hydrogen ion. These protons (H^+) actually dissolve into the surrounding metal walls. They get trapped at imperfections in the metallic structures, combine to form molecular hydrogen (H_2), and in so doing, create localized hydrogen pressures or stresses inside a metal wall. This is the origin of hydrogen-assisted stress corrosion cracking, which is a leading cause of vessel failure and death and disaster in refineries and chemical plants.

$Fe(Cl)_2$ is water-soluble. As the HCl produced is present in only a few ppm, it does not directly result in very much metal loss, except when aided by H_2S in its evil effects.

HCl is a very strong acid. H_2S is a much weaker, less reactive acid. Ordinarily, a weak acid like H_2S could never displace from its salt (i.e., $FeCl_2$) a strong acid like HCl. However, the concentration of H_2S is typically 10,000 times greater in a refinery environment than that of HCl. Thus, the following unfortunate reaction will take place:

- $H_2S + FeCl_2 = Fe(HS) + HCl$ (4)

Again, this reaction can only proceed in an aqueous phase. The resulting iron sulfide, which is insoluble in water, is that black, thin powder we find all over the plant. The HCl so liberated from the iron sulfide goes on to destroy a new molecule of iron.

It's rather like hydrochloric acid is a "catalyst for corrosion," when mixed with water and high concentrations of hydrogen sulfide.

Preventing Corrosion

In refineries, most of the corrosion with the origin I have just described can be mitigated by improved desalter operation to enhance the $Mg(Cl)_2$ extraction. In order of importance, here are steps to minimize $MgCl_2$ escape from the desalter:

1. Double desalting.
2. Inject all wash water upstream of the first preheat exchanger.
3. Use 6 wt% or more water on crude; 4% is too little.
4. For heavier crudes, use desalter temperatures of 280°–300°F. For lighter crude, use 250°F–270°F.
5. If the sour water stripper bottoms is used as desalter wash water, keep NH_3 concentration at less than 20 ppm.
6. Control desalter pH to 6.5–7.5.
7. Continuous use of water recirculation through bottom mud wash nozzles to reduce solids buildup is helpful, but promotes sludge carryover to the exchanger train downstream of the desalter.
8. Add 50% of chemical demulsifiers to the unit charge tank, and add the rest to the desalter itself.
9. Use a conductivity probe to monitor the oil-to-emulsion interface levels.
10. BS&W should be NMT 0.2–0.3 wt.% in the desalter crude effluent stream.

Reformer Hydrogen

Chlorides escape from the naphtha reformer with the surplus hydrogen used in the downstream hydrotreating units. Formerly, we used a 10% caustic wash to extract HCl from the reformer off-gas. Current practice is to remove HCl from the hydrogen off-gas with a molecular sieve regenerable absorption tower.

Regardless of the method used, my clients often neglect monitoring of their chloride removal from reformer hydrogen off-gas, with sometimes catastrophic downstream results due to equation (4), above. Periodic checking with a Drager tube of the hydrogen off-gas for chloride concentration may be an effective way to minimize disaster in the refinery downstream hydroprocessing units.

Carbonic Acid

An excellent way to remove rusty stains from your toilet bowl is to pour a can of Coke directly over the rust. It's the dissolved CO_2 in the aqueous phase (H_2CO_3), carbonic acid, that attacks the iron deposits in your toilet. Carbonic acid, unlike HCl and sulfuric acid (H_2SO_4), is quite corrosive to carbon steel, even at a relatively high pH. If water contaminated with hydrochloric acid has a pH of 5½ to 6, that is not too alarming. However, water with the same pH, due to carbonic acid contamination, will be quite corrosive.

I normally encounter corrosive carbon dioxide (CO_2) in an aqueous phase in steam condensate. The CO_2 originates in steam due to calcium carbonates (Ca_2CO_3) in the steam generator boiler feed water supply. Upon heating in the boiler, the carbonates dissociate into:

- $Ca_2CO_3 + H_2O = Ca(OH)_2 + CO_2$

The CO_2 is a non-condensable component, left behind in heat exchangers, after the steam condenses. As a non-condensable, it accumulates inside of the steam reboilers and the steam shell-in-tube heat exchangers. If not vented off, or neutralized, it will form high concentrations of carbonic acid in the exchanger and become very corrosive to carbon steel.

If the steam is on the tube side of a horizontal shell and tube exchanger (Fig. 24–1), then the accumulated carbon dioxide must be vented off beneath the channel head pass partition baffle (point "A" in Fig. 24–1). Venting from the high-point connection (point "B") will only vent steam and not remove any of the accumulated carbon dioxide. My practice is to vent continuously from point "A" through a restriction orifice, sized of 0.5% of the total supply of the steam flow. Actually, this is a Shell Oil standard I use.

At Texas City, we handled this accumulation of corrosive carbonic acid in the channel head of our steam heated reboilers by injecting neutralizing amines into the reboiler's steam supply. This was done because the plant manager was treated to hunting trips by the chemical vendors.

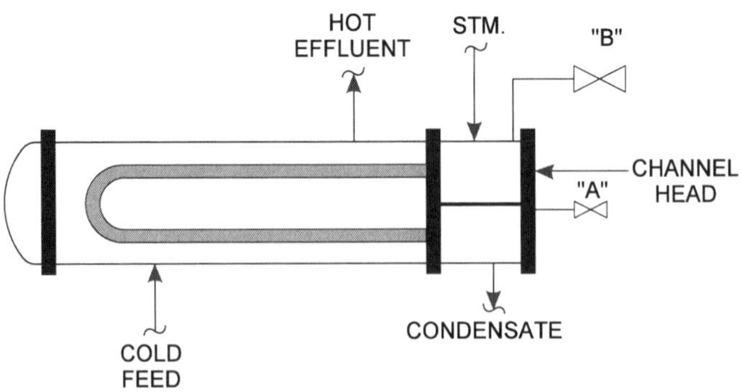

Fig. 24–1. Correct CO_2 vent is thru PT. A- NOT PT. B to prevent carbonic acid corrosion and tube failures.

Other Corrosive Agents

Naphthenic acids and sulfur compounds may also prove to be active corrosive agents in refineries. However, high-corrosion rates for these elements often require higher temperatures and/or high velocities. In general, most of the corrosion problems I have encountered as an operating superintendent and process design engineer have been associated with aqueous phases. In general, water is the most corrosive medium in many applications, especially when combined with NH_4Cl salt deposits and hydrogen sulfide vapors.

Significant Corrosion Failures

In my long career, I have been personally involved in a number of refinery catastrophic failures caused by corrosion:

- Unocal, Chicago. LPG absorber—hydrogen-assisted weld failure: 17 people in the fire crew killed.
- Shell, Norco. Cat cracker fractionator blown off its foundation: 6 operators killed.
- American Oil, Texas City. Alkylation unit with H_2SO_4 acid corrosion: world's largest gasoline production shutdown for several weeks, but no one hurt.

Without exception, these failures—and every one of the many such incidents I have been involved in during the past 58 years—could have been prevented with little or no cost, by applying readily available process engineering technology. The incident cited above at Texas City, I personally caused as a consequence of lack of experience.

Lessons Learned

- Corrosion typically requires (at low temperatures) an aqueous phase.
- Corrosion creates the environment for hydrogen-assisted stress corrosion cracking.
- Ionic hydrogen can, and has, caused vessels to fail—in hours, not years—and has taken many lives in the process industry.
- Water is the basis for most corrosion mechanisms in process units. Keep your drains operational.

25

Vessel Failure Due to Hydrogen Attack

Introduction

This chapter is a detailed description of the C_3-C_4 liquid-liquid extraction tower at the Unocal Refinery in Lemont, Illinois. I was the only outside expert witness for this disaster that killed 17 plant personnel. This incident is the prime example in the refinery industry of hydrogen-assisted stress corrosion cracking, which illustrated the need for post-weld heat-treating of vessels and piping in amine service. This is now an industry standard.

Purpose

Prior to this incident in Lemont, vessels in rich amine service that were used to absorb H_2S from hydrocarbons were not considered to be in hydrogen service and were therefore not exposed to hydrogen-assisted stress corrosion cracking. Therefore, the engineering company (which was the leading designer of refinery process equipment in the US) did not specify that the amine C_3-C_4 liquid-liquid absorber be post-weld heat-treated. Based, to some extent, on my expert opinion filed with the court, vessels in amine service are now all post-weld heat-treated.

I grew up in Brooklyn, New York, and lived in an apartment on the fifth floor. It was heated with steam radiators. The low-pressure steam, generated in a coal-fired boiler in the basement, rose up 50 feet to the cast iron radiator in my bedroom. There, the steam condensed to water. Latent heat was liberated, as the steam condensed and cooled.

As it condensed, the steam—at a pressure perhaps slightly more than 14.7 psia (atmospheric pressure)—liberated about 1,000 BTUs per pound. The steam condensate then cooled from 212°F to around 100°F. As the specific heat of water is:

- 1.0 BTU/LB/°F,

the heat liberated due to cooling the steam condensate was:

- (212 – 100) • (1.0) = 112 BTU/LB.

So each pound of steam provided:

- 1,000 + 112 = 1,112 BTU/HR

of heat to my small bedroom. That is, about 90% of the heat liberated in the radiator came from latent heat, rather than from sensible heat.

The steam condensate drained back through a condensate return line to the boiler in the basement. There, Mr. Charles shoveled the shiny black coal into the boiler's glowing firebox.

Non-Condensable Vent

On the upper portion of the steam radiator in my bedroom, there was a small metal bottle with a tiny valve on its top end. "What is that for?" I wondered for many years. One day, my mom, who had noticed that my bedroom was getting progressively more chilly, complained to the building's superintendent.

Later that day, Mr. Charles came into my room to "bleed off the air" trapped inside the radiator. "Sonny," he explained, "that air gets trapped inside the top of that radiator. It stops the flow of the steam from the boiler."

Pointing at the mystery metal bottle on the top of the radiator, Mr. Charles explained, "That's the non-condensable vent. If I open this little valve, then the non-condensable air will blow out and make room for more steam to flow into your radiator."

"But how does the air get into the radiator in the first place?" I asked. "Isn't the steam pressure too high for this to happen?"

"Well, sonny, it ain't actually air. I'll show you." He then opened the small non-condensable vent valve and touched a lit match to the evolved vapor—and it burned!

"Why is that air burning?" I asked.

"'Cause it ain't air. It's something else. It always happens that way. It's some kind of gas that forms inside the radiator coil," explained Mr. Charles.

Many years later, when I had occasions to interact with plumbers, they all knew that the combustible gas being vented from the radiator was actually HYDROGEN! But what was the origin of this hydrogen?

Origin of Hydrogen Inside Vessels and Heat Exchangers

Coca-Cola is quite corrosive. It's full of the two main ingredients that cause iron to corrode: CO_2 and H_2O. These, taken together, form carbonic acid (H_2CO_3). When carbonic acid (i.e., carbon dioxide dissolved in water) interacts with iron, the result is:

- $Fe + H_2CO_3 \rightarrow FeHCO_3$ (1)

But there is something missing from the above equation. I have left out an H^+ (a hydrogen ion) or proton.

All aqueous-phase corrosion always produces ionic hydrogen plus a salt, such as rust or iron carbonate, as shown in equation (1). These hydrogen ions normally join up with a sister ion to produce molecular hydrogen. That's the stuff Mr. Charles ignited when he vented my steam radiator in New York. Hydrogen is just as flammable as the natural gas you burn in your home.

But these hydrogen ions may also diffuse into the molecular structure of steel pipes and vessels. This is not normally a problem. The ionic hydrogen simply passes through the molecular structure of the steel vessel wall and combines into molecular hydrogen at the exterior of the vessel.

Except if there is an imperfection in the vessel wall, because the vessel has not been "post-weld heat-treated" to remove the stresses due to careless welding. Then, the molecules of hydrogen formed from the protons (i.e., ionic hydrogen) become trapped in the tiny cracks in the heat-affected zones of the welds. The molecular hydrogen exerts greater and greater pressures, until the weld cracks. This is called "hydrogen-assisted stress corrosion cracking," which is one of the leading causes of process plant fires, explosions, disasters, and deaths.

Hydrogen cyanides (HCN) formed in an FCU tend to retard the combination of ionic hydrogen inside a vessel into molecular hydrogen, thus promoting the penetration of ionic hydrogen (i.e., protons) into vessel walls. That is why it's so important to x-ray welds on piping and vessels in FCU service and to post-weld such welds.

Examples of Hydrogen-Assisted Stress Corrosion Cracking

Here are a few of the examples of this problem that I recall from my personal history:

- Unocal, Chicago—Failure of a liquid-liquid C_3-C_4 amine absorber—described above
- American Oil, Texas City—Piping failure of the effluent treatment piping in a sulfuric acid alkylation unit
- Chevron, Salt Lake City—De-ethanizer cracking in a fluid cat cracker; vessel replaced.

Only too often, engineers and operators think that a vessel cannot suffer from hydrogen-assisted stress corrosion cracking because no hydrogen is present. But if there is aqueous corrosion and moisture, then ionic hydrogen will be certain to be present.

Lessons Learned

Amine systems are all potentially corrosive. One product of corrosion is always hydrogen. If a weld has not been post-weld heat-treated to relieve stresses caused by poor welding technique, then hydrogen ions get trapped inside the weld. This, sooner or later, may cause the weld to crack. Once a weld begins to crack, metal failure will follow in a matter of minutes or hours.

The quality of the circulating lean amine needs to be monitored for the accumulation of heat-stable salts. Above 3 wt.% heat-stable salts, the amine should be "Reclaimed" to remove these salts. The amine acid gas loadings should not become excessive (i.e., NMT 0.45 moles of H_2S + CO_2 ÷ moles of amine).

Do not expect a contractor doing a repair weld on a vessel to decide if post-weld heat-treatment is required—that's our responsibility.

26

Energy Release by Combustion

Introduction

In most plants, fuel is consumed in three types of process applications:

- Gas-fired turbines
- Reciprocating engines
- Fired heaters

Purpose

Engine horsepower output can be influenced by factors other than the fuel consumption rates. These other factors include ambient air temperatures, elevation, relative humidity of the air, and combustion air supply pressure (i.e., from a natural gas-fired turbine).

The amount of energy that can be derived by the combustion of a hydrocarbon fuel is proportional to the (a) percent of carbon in the fuel, (b) percent of hydrogen in the fuel, and (c) properties of the combustion air. It is this last factor—the properties of the combustion air—that is discussed in this chapter. Not the air's oxygen content, which is invariant, but the air's:

- Temperature
- Pressure
- Humidity.

These factors can have a relatively large effect on the performance of many types of process equipment.

Engine Horsepower

The horsepower developed by any engine driven by combustion is proportional to only one factor—air! The greater the air supply to your car's engine, the more horsepower it will develop. A supercharger, or turbocharger, is just a way to force more air into the engine's cylinder. Pumping in more fuel does not develop more horsepower, unless the airflow can also be increased.

I won the annual half-marathon race in New Orleans (in the 70+ age group). Not because I consumed a lot of candy, but because I run every day and have developed good lung capacity. Next year, I'm hoping to win the New Orleans Marathon in the 80+ age group. My actual running time, however, will depend to a great extent on the ambient temperature. Every 5°F reduction in temperature should reduce my running time by 1%.

Air Temperature Affects Horsepower

When I run in New Orleans, I run slower than in Alberta, Northern Canada. I generate less horsepower because when I run at home, the air temperature is often close to 98°F. In Canada, it's typically 40°. Hotter air is less dense. That's why hot air balloons rise.

But how much less dense is the hot air than the cold air? Let's first convert to degrees Rankine (or Kelvin, if we are in Europe).

Table 26–1. Converting Degrees

USA - Rankine	Europe - Kelvin
0°F = 460°R	0°C = 273°K
40°F = 500°R	20°C = 293°K
100°F = 560°R	60°C = 333°K

(°R or °K is called "absolute temperature.")

The density of a gas is inversely proportional to the absolute (i.e., °R or °K) temperature. Therefore, 100°F air is less dense than 40° air:

- 500°R/560°R = 89%

(100°F = 560°R) and (40°F = 500°R)

So, just based on the extra 60°F, I would lose 11% of my body's power output (100% - 89% = 11%).

Effect of Elevation

When I visit Northern Alberta in Fort McMurray, Canada, the plant I work in (Suncor) is at an elevation of 3,000 feet. Atmospheric pressure is consequently lower than in New Orleans:

- Alberta—13.2 psia
- New Orleans—14.7 psia

Air density in New Orleans, due to its sea level location, is 10% greater than in Northern Alberta. This effect is additive to the effect of temperature if we live at a low elevation in a cold climate:

- (14.7 psia – 13.2 psia) ÷ 14.7 psia = 10%

Effect of Moisture

In the winter, the air in Northern Alberta, in Fort McMurray, at the Suncor Oil Sands Plant, is bone dry, with 0% relative humidity. However, in New Orleans, the relative humidity is often close to 100%. That is, the air is saturated with water. If the air temperature is 100°F, and the relative humidity is 100%, then, using the following Table 26-2, we can make some further calculations to assess the effect of moisture in the air.

Table 26–2. Effect of moisture in the air

Temperature, °F	Vapor Pressure of Water, psia
32	0.09
40	0.25
60	0.52
80	1.03
100	1.94

We can calculate that the concentration of moisture in the air in New Orleans is:

- 1.94/14.7 = 13%.

Therefore, the tremendous humidity in my hometown in Louisiana, in mid-summer, will reduce the concentration of oxygen in the air, per cubic foot, by 13%.

So, if you drove your car at a high elevation, a high-ambient temperature, and at 100% relative humidity, then your engine's horsepower would diminish by:

- (89%) (90%) (100% - 13%)
- Effect of (temperature) (elevation) (humidity) = 71%.

That is, the combined effects of the 100°F air, perhaps 4,000 ft. of elevation, plus 100% relative humidity, could reduce your car engine's horsepower output by 29%! Typically, the effects of altered elevation are most evident.

From my perspective as a runner, when I run in Denver, I run a lot slower, as I also do in New Orleans in the summer. When I operated a gas turbine in Laredo, Texas, the machine would slow down by about 5% when ambient temperatures increased during the course of the day, from 60°F at 6:00 a.m. to 105°F by 5:00 p.m.

Compressing Combustion Air

The front end of a jet engine is a multi-stage centrifugal air compressor. The compressed air then flows to the combustion zone, where turbine fuel is injected. The heat liberated drives a rotor, which spins the multi-stage centrifugal compressor.

The supercharger on a car motor is an air compressor. It pushes extra air into the car's cylinders. A turbocharger does the same thing. The turbocharger is driven from an engine drive belt. Use of oxygen-enriched air is one method to expand the capacity of any turbine or internal combustion engine. Liquid oxygen was used on the rockets developed by the Germans in World War II (i.e., the V-2s), in 1945.

Fired Heaters

If we extrapolate from the internal combustion engines to fired heaters, we'll see the same problem. The pounds of combustion air will diminish to the burners if the ambient conditions become more humid and hotter. Also, the hotter air reduces the "stack draft." But that's a story for another time.

It's a good thing our little planet is currently surrounded by a blanket of air with 21% oxygen provided by plants. Otherwise, we would really be in serious trouble regarding combustion applications. Originally, the oxygen content of our atmosphere, was 0.3%.

Natural Gas-Fired Compressors

In a jet engine, the gas turbine exhaust provides the thrust that drives the plane and also drives the three-stage centrifugal combustion air compressor. In a natural gas-fired centrifugal compressor, about 40% of the energy produced from combusting the fuel gas drives the three-stage centrifugal combustion air compressor. About 30% of the energy produced from combusting the fuel gas drives the natural gas multi-stage centrifugal compressor, and about 30% of the energy produced from combusting the fuel gas blows out of the exhaust vent at 1000°F into the atmosphere (as was the case south of Laredo, Texas, on my natural gas compression unit in 1986).

I was always annoyed in Laredo when my gas compressors would slow down in the late afternoon, just when I needed their speed the most, to maximize delivery of natural gas to Exxon, who supplied the natural gas to utilities generating electric power for homeowners in New York and Chicago.

Lessons Learned

As described above, in Laredo, Texas, I was surprised to find my giant natural gas compressor, driven by a 10,000 BHP gas turbine, would progressively lose horsepower as the ambient air temperature increased from 40°F to 100°F between 6:00 a.m. and 5:00 p.m. This is a good example as to how a personal experience can aid in understanding of a general process application. I had never thought before, as to how ambient conditions can affect engine horsepower output.

27

Case Histories: Avoiding Death and Disaster in Process Units

Introduction

There are a number of sources for the information presented in this chapter. One source is my testimony as an expert witness for contractors that were involved in litigation as a result of injuries and/or fatalities in refinery process units. Also, I was an expert witness for litigants suing major refineries for personnel injury and/or property damage. Other sources of information are my experience as a technical manager and plant manager of the Good Hope Refinery in southern Louisiana, as well as my experience as a young engineer in charge of the world's largest H_2SO_4 alkylation unit.

Purpose

Refineries are dangerous places, where hydrocarbons, air, fire, and H_2S are mixed together in a controlled manner. In the stories related in this chapter, this mixing became uncontrolled. The following are keys to safety in process plant operations:

- Do the operators understand the process?
- Do the operators understand the instrumentation?
- Is the engineering staff working closely with the hourly operators to fulfill the first two items above?
- Did the original engineering design staff evaluate potential process safety hazards in the course of their process design work?

Raffinate Splitter Explosion

Seventeen people were killed, hundreds injured, and 5 billion dollars lost when a gasoline raffinate splitter blew out a cloud of gasoline, which exploded at a refinery

I helped design and operate. It's true that there were a number of design problems that contributed to the catastrophe (Fig. 27–1):

- The relief valve was on the vapor line located 70 feet below the top head of the splitter tower.
- The relief valve opened into an atmospheric vent (i.e., the blowdown stack), rather than into the flare system.
- Liquid drained from the atmospheric vent (i.e., blowdown drum—see Fig. 27–1) through a loop seal, which vapor-locked.

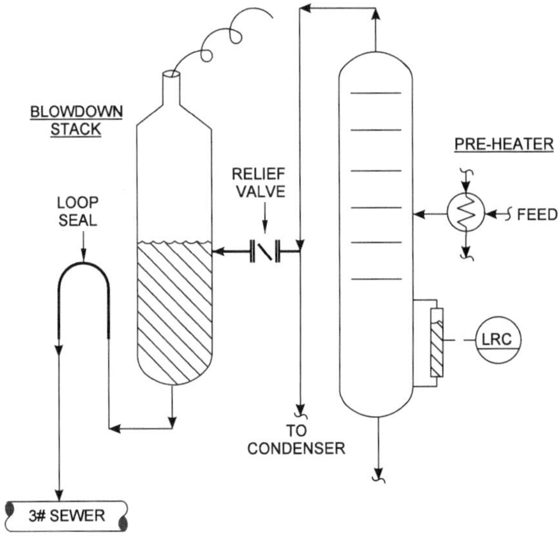

Fig. 27–1. Raffinate splitter-16 killed

But the real cause of the problem was that the operators didn't understand how their level indication worked. Some of the operators thought they might have a very high-liquid level and that the tower was full. But when they increased the feed temperature, they noted a reduction in the bottom's level indication on their control panel. They then concluded that the tower's level indication was okay and continued to bring more feed into the splitter. The tower then overflowed and pressured-up the relief valve, due to the weight of gasoline (i.e., the hydrostatic head of liquid above the relief valve) overflowing the splitter and pressing down on the relief valve.

The two factors that operators missed were

1. Hotter gasoline is less dense.
2. The level indicator will read a reduced level when the density of the gasoline is reduced, even though the real level is actually more than 100% of the level indication range.

I learned about this when my Dad taught me to check my car's engine oil. The oil level ran 1–2 inches higher on the dipstick when the engine was hot.

LPG Amine Extraction Tower Failure

Seventeen volunteers of the fire response team and one operator were killed at the refinery, and the alky unit was destroyed, when a C_3-C_4 LPG MEA (monoethanolamine) H_2S extraction tower failed. The vessel ruptured along a circumferential repair weld.

It is, in fact, true that the refiner had been quite negligent regarding taking measures to avoid the explosion. These were some of the errors made:

- Used MEA, rather than less aggressive DEA or MDEA.
- Failed to run the MEA reclaimer to remove heat-stable salts from the circulating amine. The heat-stable salts promote corrosion.
- Allowed excessive acid gas loading of the MEA solution, which also accelerated corrosion.
- Failed to properly inspect the vessel for cracks at repair welds during turnarounds (Fig. 27–2).

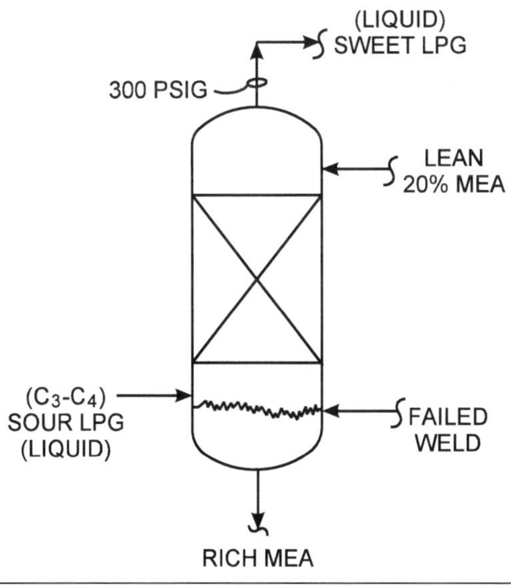

Fig. 27–2. Lack of post-weld heat-treating results in death and disaster.

But the fundamental cause of the failure was a bad decision that occurred before the vessel was fabricated. The designer had failed to specify on the vessel sketch that the extraction tower must be ***post-weld heat-treated to relieve stresses in the heat-affected zones of welds***.

Thus, when a repair weld was made many years later, the repair weld was not post-weld heat-treated either. What the engineering design contractor had missed was that:

- Amine treating of H_2S extraction is a potentially corrosive service.
- MEA is more corrosive than DEA or MDEA.
- When steel corrodes, one product of corrosion is iron sulfide.
- The other product of corrosion is a hydrogen ion.

Ionic hydrogen will penetrate welds and accumulate as molecular hydrogen in unstressed relieved welds and cause the welds to fail. Neither the refiner nor the engineering contractor was found liable for this explosion. But my client, who did the repair weld, was found, quite unjustly, to be at fault.

Failure of Catalytic Cracking Unit

In the late 1980s, an explosion at a FCU blew the large catalytic cracker fractionator over. It landed on the control room and killed all six of the operators inside.

Twenty-five thousand local residents then sued the refiner for $1.3 billion dollars for damages to their property. The refiner had responsibly and promptly paid for all legitimate claims. After two years of litigation, the Plaintiffs' Legal Committee (a group of seven prominent attorneys) settled with the refiner for $400,000,000 (a compromise from their goal of $1,300,000,000). The lawyers' share was $160,000,000. The litigants received $240,000,000. I was paid $9,800 as the Plaintiffs' Legal Committee's expert witness.

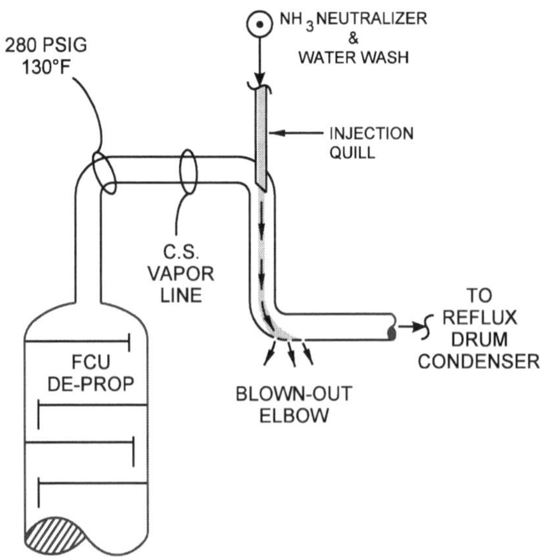

Fig. 27–3. Fluid catalytic cracker unit vapor line failure due to quill location.

What Happened?

An overhead 6-inch carbon-steel vapor line from the cracking unit depropanizer failed. A free air vapor cloud of propane-propylene detonated and blew the main fractionator (perhaps 16' I.D. x 100' T-T) off its foundation and killed six operators in the control room.

I visited the control center about five years ago. They have a nice plaque commemorating their fallen comrades—but no factual explanation as to what mistakes were made, that caused this incident and the loss of life.

It could be that I am the only person still active in the refining industry who actually understood the mistakes that were made, which were a prerequisite for the failure of the overhead vapor line.

A Litany of Errors

Let me describe what happened. The depropanizer overhead hydrocarbon vapor consisted of the following approximate mixture:

- A few % ethane
- A few % H_2S
- A few % butanes
- About 40% propane
- About 50% propylene
- Saturated with water

The overhead vapors were at a pressure of roughly 280 psig and 130°F.

All of the above is typical for this service. What was not typical was that the water drawn from the reflux drum boot was acidic. Typically, the NH_3 content of the fluid cracker depropanizer reflux drum boot would be sufficient to result in a pH of 8–10. Why this was not the case here (the boot pH, prior to the addition of the neutralizer, was about 5–6 pH), I cannot say.

The refiner had found the overhead condenser's carbon-steel tubes to be subject to high rates of corrosion. Thus, the tubes were changed to alloy tubes—but not the carbon-steel vapor line upstream of the condenser.

To control corrosion, a water injection point was installed upstream of the condenser. Water was injected through a "quill." That is a piece of tubing, cut at an angle, and inserted just into the piping, upstream of an elbow (Fig. 27–4) in the vapor line, feeding the condenser.

NH_3 was added to this wash water to control the reflux drum water boot pH at about 7.5–8.0. The water was clarified river water, without deaeration.

After several years of operation, the elbow blew out, and the catastrophe I described above followed.

What Went Wrong?

I inspected the failed elbow in the investigation lab and reviewed all of the internal documents in my role as an expert witness. Here is a tabulation of the errors that I documented, in order of importance:

1. A "Bete" or "Spraying Systems" spray or mist nozzle should have been used. It was obvious from my inspection of the elbow failure pattern that the outer periphery of the water jet had absorbed most of the acidic components in the vapor. A groove had been eaten through the pipe wall exactly around the outer impingement area of the spray. Inside and outside of this ½–1-inch groove, the pipe was at its original thickness.
2. The "quill" (the cut-off tubing) was pointing directly down at the failed elbow, which was about 30 inches below. The quill should have been in a horizontal run of line.
3. If there was a need to change the condenser tubes from carbon steel to alloy tubes, why not put the carbon-steel piping upstream of the condenser as well?
4. Why were the catalytic crack depropanizer overhead vapors acidic? Typically, towers in this service have a pH of 8–10 in the reflux drum water boot. The refiner should have investigated this many years ago, at the start of operations, when they first observed the low pH problems of the water in the boot and the corrosion of the tubes in the overhead condenser.
5. The wash water should have been deaerated water, not filtered river water from the river, which contains corrosive dissolved oxygen.
6. The refiner had been advised—by their chemical treatment provider—of a potential problem in this service, a year before the failure.

Bhopal Disaster

This pesticide plant killed over 10,000 people in Bhopal, India. As a consequence, a large, famous chemical company was liquidated. Bhopal is the worst accident that has ever happened in the process industry.

The cause of this incident has been attributed to a faulty mechanical seal on the methyl isocyanate pump. The pump had been designed with a marginal available net positive suction head (NPSH), which resulted in persistent cavitation at the entrance to the pump impeller. This then caused damage to the seal and frequent seal leaks (Fig. 27–4).

The resulting methyl isocyanate leaks were an irritant, but not fatal. To avoid this relatively minor problem, plant operators devised a solution:

- The pump was shut down.
- The bypass valve around the pump was opened (Fig. 27–4).
- The isocyanate vessel was pressured-up by several bars (perhaps 20–30 psi), using an existing nitrogen purge connection.

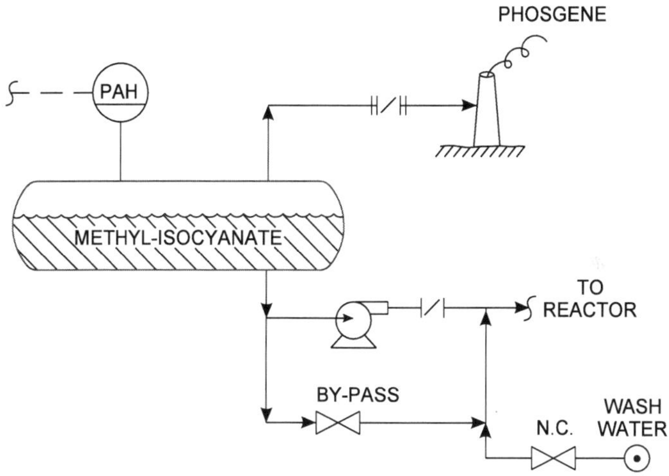

Fig. 27–4. Union Carbide Bhopal, India; 10,000 people killed

As a consequence of the isocyanate vessel operating at a higher-than-normal pressure, the vessel pressure alarm would periodically be activated. To avoid this annoyance, the alarm was deactivated.

During World War I, the Germans introduced poison gas, the most terrible being phosgene. These are the steps to produce phosgene:

- Step 1—Take methyl isocyanate.
- Step 2—Add water.
- Step 3—Mix gently for 5 minutes.

At the Bhopal plant, wash water was used periodically to flush out the feed line to the reactor. With the methyl isocyanate pump running and the pump bypass valve shut, water could never flow back into the vessel.

However, with the pump out of service and the bypass valve open, water could—and did—back up into the vessel. The result was a rapid evolution of phosgene vapors that caused the vessel to over-pressure. The relief valve opened, and fatal phosgene gas settled over the city of Bhopal, India. The water injection continued even as people died.

But why?

Because the high-pressure alarm had been deactivated, due to running the vessel at a higher-than-normal pressure with nitrogen. Thus, the control panel operator was not alerted to the problem of the excessive vessel pressure.

Death by Nitrogen

Three contractor maintenance workers died from nitrogen suffocation in a hydrocracker reactor. The reactors had been purged with nitrogen prior to issuing entry permits. Two of the three identical reactors had been tested for breathability. The third was full of nitrogen only. The three workers entered the wrong reactor, collapsed, and died.

Humans only feel difficulty in respiration if CO_2 in air exceeds 0.5%. If no CO_2 is present, regardless of the lack of O_2, no respiration discomfort will be experienced.

Lessons Learned

- Stress-relieve vessels and piping welds.
- Know when equipment has been properly depressured.
- Understand how level measurements vary with liquid density.
- Do not use wash water, without a spray nozzle, that impinges on piping.
- Know which alarms should never be deactivated.

Understand how the process equipment works and use your experience to anticipate what can go wrong when lessons learned from past experience are ignored. I have made many mistakes in my long career as a process engineer, but I like to take pride in not making the same mistake twice.

Almost all the incidents I have been involved with were caused by this combination:

- Instrumental malfunctions
- Lack of operator understanding of process safety hazards
- Basic process design errors promoting unit instability
- The corrosive effects of water, oxygen, and CO_2 that combine to generate ionic hydrogen
- Welds that have not been post-weld stress relieved

28

Basic Safety Considerations

Introduction

Process plants—and most especially petroleum refineries—are inherently dangerous facilities. Part of the reason is that we are mixing (hopefully in a controlled fashion) these sorts of substances:

- Fire
- Air
- Hydrocarbons

This mixing is often done at elevated temperatures and elevated pressures. A fluid catalytic cracker catalyst flow is at 1,000°F. Heater outlet temperatures from a delayed coker are 930°F. The tube skin temperature of a crude heater may approach 1,275°F.

Hydrocracker reactor vessels will operate at over 2,000 psig. I once designed a "white oil" hydrotreater to operate at 3,000 psig.

Purpose

The main danger that is faced in process operations is not so much a consequence of excessive temperatures and pressures. More importantly, it's a combination of three factors:

1. Intrinsic process design errors.
2. Lack of the unit engineer's identifying, and correcting, ongoing safety concerns by the operating personnel.
3. Operators never understanding how the process works, and not understanding the plant's instrumentation.

Safety on a process unit is not so much a function of attitude, alertness, or awareness. No one I have met in a refinery wants to work in a hazardous environment. Process people are hired because they are responsible, smart, and careful. Safety is largely a function of what operators and engineers know about the processes we work with. It's knowledge that prevents accidents and injuries.

Autoignition

Autoignition refers to the temperature at which a dry material will start to burn when exposed to air without an external source of ignition. Question: which has a lower autoignition temperature—gasoline or asphalt?

The answer is asphalt. Vacuum resid autoignition temperature is around 300°F. Not to say that if you draw a sample of asphalt into a can, it will start to burn at 400°F. It will just smoke. But if you spill the contents of the can on a large, porous surface like your clothes, it will autoignite.

Gasoline, on the other hand, has an autoignition temperature in excess of 450°F. The higher the octane, the higher the autoignition temperature. Hydrogen's theoretical autoignition temperature is very high, but in practice, it ignites relatively easily.

Nitrogen

In 1982, I connected an N_2 line to instrument air, as a backup to our unreliable instrument air compressor at the Good Hope Refinery. When the N_2 backup opened to offset the failure of the air compressor, air in the control center (which was used in our pneumatic process control panel instruments) was gradually displaced with nitrogen. I was sitting in the control room when the low-oxygen alarm sounded at 18%. No one experienced any discomfort or shortness of breath. That discomfort is due to excess CO_2 (0.5%) and not lack of oxygen in the air we breathe.

The first sign of low O_2 levels is death. In the 1960s, in Texas City, three workers who entered a hydrocracker reactor during a turnaround that had been purged with nitrogen were killed (see previous chapter).

Hydrogen Sulfide

Hydrogen sulfide (H_2S) is not dangerous, as long as you can smell it. In fatal concentrations of 1,000 ppm or even lower, the first whiff of H_2S deadens one's olfactory nerves. The next few breaths lead to unconsciousness. Then, one just lies there and dies. We killed three workers in Texas City, on #2 Alky, when H_2S backed out of a flare line. The H_2S feed to a sulfuric acid plant had been diverted to the flare. Unfortunately, the flare connection at the alky unit had not been blinded-off. A serious safety omission. Always blind-off the flare connection during a unit turnaround.

Pyrophoric Iron

Iron sulfide is a black, finely divided solid that looks like coke. As long as it is wet, it is quite inert. However, when iron sulfide dries out, it ignites at ambient temperatures.

I recall discarding a large quantity of paper filters covered with iron sulfide into a 55-gallon drum in Texas City from our amine filter during the 1980 strike. Nothing happened for a month. Then, one warm summer day, white smoke started to flow from the drum. The iron sulfide had autoignited upon drying.

The white smoke was sulfur dioxide, which is quite as deadly as H_2S. I recall inspecting a sulfur plant condenser once, when the iron sulfide autoignited at 80°F. It burned pale blue and did not emit much heat. Luckily, I was able to escape quickly from the deadly evolved sulfur dioxide fumes, which are just as dangerous as H_2S in a confined space.

That's why we chemically clean vessels with a Zyme solution, before issuing entry permits in refineries, to remove iron sulfide deposits.

Level Indication

Liquid levels in vessels are almost always higher than they appear externally. The operators at a refinery in Texas City missed this point and then caused the raffinate splitter explosion that killed 17 contractors.

In this case, the density of the liquid in the tower, due to the naphtha startup composition being lighter than the normal liquid in the tower bottoms, was lower than normal. Hence, the indicated level in the control room was less than the actual level. The problem is that we measure DP across level taps (Fig. 28–1), not the actual liquid levels.

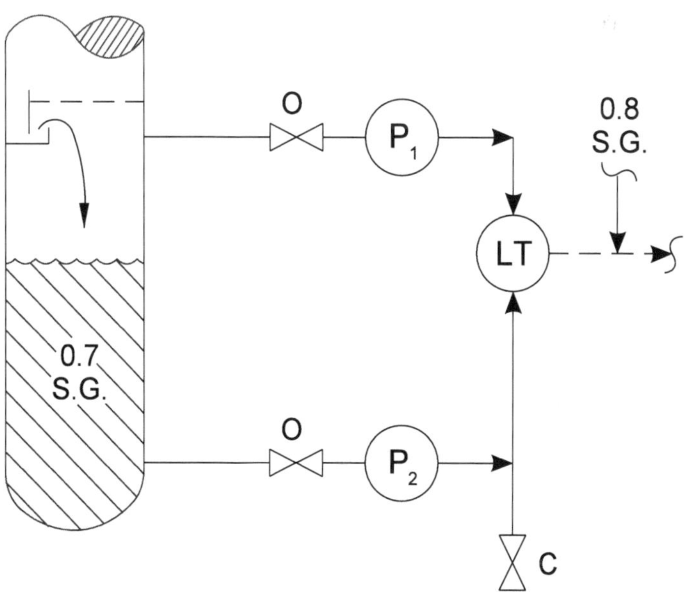

Fig. 28–1. DP÷S.G. Used to calculate level.

The following are some other causes of a lower-than-actual level indication:

- Foam inside vessel
- External level indication being colder than internal level, which is hotter and less dense (5% for every increase of 100°F)
- Water in the external indicator and oil-water phases inside the tower
- Stuck level ball-check valve (i.e., the Jurgenson valve)
- Too high S.G. used in level calibration by the instrument technician

Unfortunately, the use of a radiation level detector usually creates more safety problems than it solves.

Flooded Firebox

Attempts to save energy by optimizing excess O_2 can cause a heater to blow up. This happened at the Good Hope Refinery. Two operators were killed when the heater floor blew down.

If a heater firebox becomes air-deficient, further increase in fuel lowers the process outlet temperature. This can occur between 0.5% and 6% excess O_2, depending on the air-fuel mixing efficiency. If a heater is air-deficient and is run on TRC (temperature control), then the following sequence results:

- The process outlet temperature declines.
- The fuel valve opens.
- This drives the outlet temperature down further.
- That increases fuel flow.
- Black smoke will eventually be emitted from the stack.

If the operator attempts to rectify the shortage of air by opening the air registers and/or the stack damper, heaters have blown up and killed operators. I once almost killed myself on an incinerator in Texas City, by rapidly opening the incinerator's air register. The incinerator's temperature had been declining, as the panel operator was increasing the incinerator's fuel gas rate. I was blown off my feet by the small explosion.

Steam Turbine Over-Pressure

Sometimes the sequence in which we perform certain tasks is critical—starting a steam turbine being a case in point. The turbine case is not designed to withstand the motive steam pressure, but only the turbine's exhaust steam pressure.

I often forget this and open the steam supply valve first. This causes a vent on the turbine case to open and results in a loud whistling sound. This vent is not a pressure-relief valve, as many operators believe. It's the mechanism to alert the operator that he needs to immediately line up the steam exhaust valve, to avoid over-pressuring and cracking the turbine case.

Steam Turbine Over-Speed Trip

A contractor was seriously injured at a refinery in Corpus Christi when a steam turbine over-sped and flew apart. I had warned the plant manager that the over-speed trip was "wired-up" (i.e., disabled), but he ignored my warnings.

The operators had wired-up the trip mechanism because they said that the turbine would trip off too frequently because the governor speed control mechanism was not working properly. While the operators were correct, what should have been done is that the governor should have been fixed, rather than disable the trip with a piece of wire. The operators did not really understand the relationship between the governor speed controller and the over-speed trip.

Water and Corrosion

The corrosive aspect of water is often not appreciated by refinery operators. In 1976, a major explosion occurred at the Amoco Refinery in Texas City, because the operators did not consistently drain accumulated water from a sulfuric acid alkylation unit depropanizer reflux drum boot on startup.

Acid-containing components (like SO_3) mixed with the water to form an acidic aqueous phase in the reflux drum, which rapidly corroded an outlet 4-inch elbow. The failure of the elbow emitted a cloud of white propane vapor that resulted in a giant, destructive detonation. I recall the chief operator explaining to me, "But, Mr. Norm, it was just water."

Color as a Temperature Indication

I was once starting up a sulfur recovery plant, when a young engineer pointed out to me that the interior of the thermal reactor was glowing a bright orange-yellowish color.

I checked the local TI and it indicated an acceptable 1200°F. So I decided to ignore the color (which indicated a temperature of over 2500°F) and proceeded with the startup. The over-rapid heating resulted in refractory failure. A potentially catastrophic event on a sulfur plant thermal reactor.

The lesson is "color counts." When a temperature indicator contradicts a tube or refractory color in a fired heater, the operators must be guided by the color and be trained to ignore the indicated temperature on their control console. Or, better yet, use an optical pyrometer. Color does not depend on the composition of the material that is glowing.

Steam Condensate Drainage

When steam condensate does not have sufficient pressure to flow into a condensate collection line, or a steam trap malfunctions, operators will drain condensate to the "ditch" (Fig. 28–2).

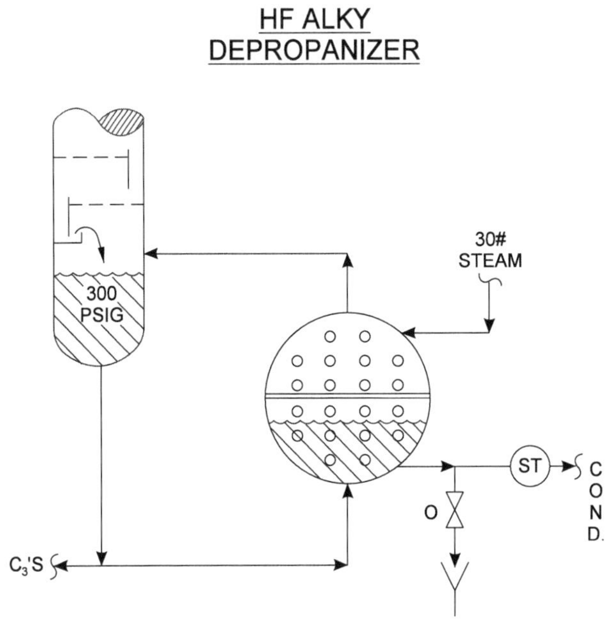

Fig. 28–2. Tube leak causes propane to sewer.

But suppose the steam is being used to reboil a 300 psig depropanizer, and a reboiler tube leak develops? Then hydrocarbons will be blowing into the ditch along with the hot water.

When operators drain steam condensate to grade from a high-pressure reboiler in hydrocarbon service, this should be viewed as a potential source of an explosive vapor cloud. At one location in Pennsylvania, I recently had a hydrocarbon vapor sensor installed with an alarm that would sound in the control center. In this case, it was not only propane that was being reboiled, but hydrofluoric acid from an HF alky unit.

I've drained steam condensate myself to the sewer from high-pressure debutanizers and depropanizers for over 50 years and never stopped to consider the potential hazards. But now I do. Never too old to learn!

Compressor Surge

A centrifugal compressor that is surging is dangerous. A piece of the rotor can break off, blow through the compressor's case, and kill someone. It's happened before.

Most operators do not know how to respond when faced with compressor surge on a motor-driven (i.e., fixed-speed) machine. Here are the most effective responses to stop compressor surge:

a. Lower the compressor discharge pressure, even if it requires venting to the flare. This is my normal response.
b. Raise suction pressure.
c. Increase the molecular weight of the gas being compressed. This is the preferred response.
d. Cool off the gas—but not if this reduces the gas molecular weight excessively.
e. Increase the flow through the compressor—but not if it requires opening the spill-back. This may increase the compressor suction temperature by a lot and hence reduce the gas density. I did this at Texas City in 1975, and put a giant centrifugal compressor into surge.

Lessons Learned

Dealing with compressor surge and the other problems I've discussed above is a complex situation. The operators and engineers need to be trained as to how the instruments and the process equipment they work with actually function, so that they know how to deal with process equipment and instrument malfunctions in a safe way, before the problems are encountered in the field.

Safety in a process plant is not a matter of people's attitude. Every operator I have worked with had a reasonable attitude to plant safety. The problem is that operators often don't understand how process equipment works. Therein lies the challenge, and the opportunity, to make our plants safer. The key to safety is to learn our lessons from other people's mistakes.

The ultimate responsibility for the long-term safety of a process plant lies in the hands of the process design engineer who has conceived the process flow configuration, as documented on the unit "PFD." The PFD ought to reflect not only a computer simulation, but the process design engineer's operating and troubleshooting field experiences.

29

Process Plant Operations and Global Warming

Introduction

The CO_2 in the atmosphere when I was born was 285 ppm. When you read this chapter, it will be 420+ ppm. The surface temperature of the Earth is a function of the heat-trapping properties of CO_2 and CH_4. This is not a matter of speculation. It is just an extrapolation of the Earth's history. Sea levels will definitely rise at a rate of 0.4 inches per year, due to a combination of water expansion by heating and glacial melting. When the hydrates offshore start to melt, methane, a greenhouse gas that is 23 times more powerful than CO_2, will escalate the rate of global warming. The rate of global warming is summarized in Table 29–1. It has doubled since 1970 – 1990. Between 1920 – 1970, global temperatures have increased by 0.5°F. Between 1970 – 2020, by 1.9°F (1°C)!

Table 29–1. Global Temperature Rise

Decade	°F
1970 – 1980	0.36
1980 - 1990	0.24
1990 – 2000	0.30
2000 – 2010	0.27
2010 – 2020	0.67

Table 29–1 suggests that in spite of efforts to date, mankind is not gaining on the problem of global warming, and is apparently losing ground.

Purpose

It isn't the politicians of the world who have created the problem. It's James Watt, Tesla, Edison, Rockefeller, and Henry Ford. People like that—you and me.

Technical folks. It's too late to avoid the effects of global warming. It needed to be addressed in 1976, when I was in charge of the world's largest alky unit and made 26,000 BSD of 100-octane gasoline for American Oil in Texas City.

Since 1975, the CO_2 level in the atmosphere has been increasing by 2.5 ppm annually. At this rate, we will be at 600—650 ppm in another 40 to 50 years. Sea levels will eventually rise by 20 feet as a result (after several centuries). That's not a guess. It's based on historical records. It's not a matter of opinion; it's a matter of calculation.

The problem has not been caused by our political leadership. It's been caused by us. Especially process engineers, who have streamlined the production of gasoline, diesel, kerosene, LPG, and plastics in the past century.

I was awarded the Hydrocarbon Processing Lifetime Achievement Award in 2018, based largely on my work as a petroleum refinery process designer and plant troubleshooter. As my daughter, Irene, observed, "I guess, Dad, this means that you are more responsible for increased CO_2 emissions than anyone in the history of the world!"

"Not quite, Irene. But I'm right up there with the historical leaders who have caused global warming, and the accelerated CO_2 and CH_4 emissions culprits."

What's to Be Done?

If we, the process engineering community, do not take corrective action, then do not expect our political leaders to solve the problem. Let me give you a recent example from a plant in Arkansas.

This plant had several relatively new, large centrifugal pumps, with VFD (variable-frequency drives). Ordinarily, a 240-volt, alternating-current motor is a fixed-speed machine. Controlling the flow from a pump driven by such a motor requires a control valve on the pump discharge. The control valve wastes energy by braking the flow.

To avoid this, a modern pump is driven by a VFD motor. No control valve is provided at the pump discharge. In this Arkansas plant, not only was a control valve provided on the discharge of the variable-speed pump, but the control valve was running 50% open. Plus, the pump speed was set at 3600 rpm!

Two years ago, in Salt Lake City, I saw the exact same scenario on a naphtha reformer charge pump, in a refinery that was suffering from an electrical power generation limit! That is, reformer capacity was curtailed by availability of electricity!

Correct Use of VFD Motors: Driving Centrifugal Pumps

1. Install a full-size impeller.
2. Install a motor consistent with the maximum impeller size and with the maximum specific gravity of the liquid to be pumped.
3. Do not provide control valves on the pump discharge.
4. Control the required variable (level, flow, pressure, temperature, etc.) by varying the speed of the motor on closed-loop control to hold the desired set-point.

After all, when you drive your car, you do not put one foot on the gas pedal and depress it to the floor and then moderate the car's speed with the other foot on the brake. Reducing speed by 10% saves about 27% of the motor's amps. That is, work varies with speed, cubed (third exponent), as per the "affinity law." I've recently seen all this working properly on a crude unit in southern Indiana.

In practice, VFD on a new pump installation, may at present, only be economically justified for motor sizes over 200 horsepower.

Excess Heater O_2

I participated in a teleconference meeting just yesterday. Management was discussing the need for an O_2 analyzer, to optimize excess oxygen—or combustion air.

This is nonsense! To optimize excess O_2 and simultaneously minimize the fuel gas consumption, an O_2 analyzer is not needed. Simply adjust your air registers (or stack damper) to minimize fuel gas consumption, when achieving the desired process outlet temperature and required charge rate. If the O_2 analyzer is located above the convective tubes, where it will be influenced by tramp air leaks, it is an especially useless and misleading instrument.

Combining the O_2 analyzer with a combustible analyzer, both located below the convective tubes and linked together with a computer, is another matter. This apparently works quite well to control the combustion air rate. I had this explained to me at a recent seminar I was teaching in Texas, and it made good sense. But this is an expensive option requiring extensive instrument maintenance and apparently a lot of maintenance for both analyzers.

Distillation

Try to minimize tower operating pressure to save energy for two reasons:

1. Reduce reboiler temperature—that is, liquid boils at a lower temperature and at a lower pressure.
2. Enhance relative volatility (i.e., the ratio of the vapor pressure of the light key component to the heavy key component). An increase in relative volatility of 10% allows a 10% reduction in reflux rate, and hence the reboiler duty.

Of course, lower pressures will increase vertical vapor velocity and entrainment, and thus sometimes may retard tray fractionation efficiency. So, to find the optimum tower pressure, you need to experiment on the unit.

The optimum pressure can never be derived by calculation, as it depends on tray deck levelness and the weir levelness, as well as the tray deck design.

Computer modeling is always interesting, but never definitive. But plant tests are! Running a radiation scan of the trays is also a useful tool.

Steam Turbines

During the 1980 strike at the AMOCO refinery in Texas City, I was working as a scab on the sulfur and amine regeneration plants that I had designed in 1978. To impress the plant manager, Mr. Durland, I decided one evening to save 400# motive steam used by the refinery's 400#-to-100# exhaust steam turbines. To accomplish this objective, I proceeded as follows:

- Step I—I reduced the turbine speed until the downstream process control valve at the pump discharge was in a mostly open, but still controllable position.
- Step II—I then closed the steam turbine's horsepower valves one at a time (Fig. 29–1; typically there are two or three of these valves), until the governor speed control valve was in a mostly open, but still controllable position. Actually, you cannot readily see the governor's position. So what I actually did was to close the hand valves, until the pressure in the steam chest (P-1) was close to the steam supply pressure (something like 20 psi). That is, 370#–380#, vs. the 400# steam header pressure. (You can measure the steam chest pressure at the steam chest condensate drain valve.)

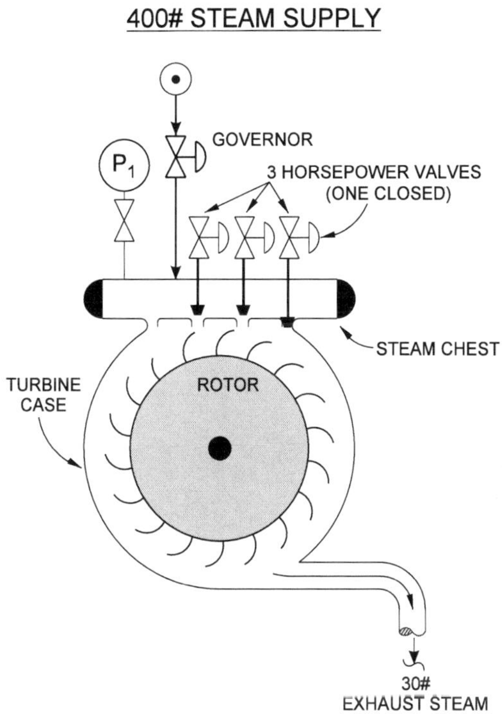

Fig. 29–1. Close horsepower valves to force governor to open and save energy.

The next day, I saw that the refinery 400# steam usage had diminished by perhaps 10%. Mr. Durland, the plant manager, did *not,* however, give me any of my anticipated and well-deserved rewards:

- A promotion
- A certificate of merit
- A gold star

Alas, when the operators came back to work after the 4½-month strike ended, they opened all the horsepower valves and reset the turbine's speed to its maximum rated 3600 rpm!

And that's the way it is. If you, the young process men and women, allow this lack of applied technology to persist for another 30–40 years, don't blame me. I'm 80 years old and certainly won't be answering my cell phone after 2040. When New Orleans and New York and Houston are periodically submerged, it will be due to a lack of applied basic process technology.

Those who do not study *geological* history are condemned to repeat it.

Lessons Learned

In this chapter, I have described a number of operational options to reduce energy consumption on process units. Converting automobiles to electric drives will reduce CO_2 emissions worldwide by roughly 15%, provided the electric power is generated by wind, water, or solar power. Improving the efficiency of steam turbines that spin rotors to produce electricity will also help. Generating power by burning natural gas rather than coal releases only half of the CO_2 per KWH. Discontinuing commercial aviation would reduce CO_2 emissions by 3%.

But apparently Mother Nature has a more drastic plan in mind to restore the equilibrium to the earth's atmosphere.

Conclusion

Nature of Process Engineering

After 58 years, I'm still filled with enthusiasm when I go to a refinery to unravel process problems. I'm still proud to tell acquaintances that I'm a Chemical Engineer.

I have been practicing Process Engineering as:

- Tech Service Engineer
- Process Design Engineer
- Operating Superintendent
- Tech Service Manager
- Field Troubleshooter
- Plant Manager
- Consultant
- Seminar Instructor

It's a wonderful job. Everyone I deal with is honest. All the operators are dedicated to making their units run better. Problems are complex, but solvable. I'm learning new things every day, and helping the society I live in function efficiently.

I was born May 7, 1942. So, as you read these words, I'll likely be 80+ years of age. Yet, I still do my own field troubleshooting and run five kilometers every day, without fail. Of course, the day will come when I can't go on. I'll forget how to do an equilibrium flash calculation or how to mentally convert moles to vapor volume. Or I won't be able to run because my left knee hurts too badly.

That day is surely coming. But each day, I say to myself, "That day is not today." Maybe it's cold and raining. Maybe a client went with a different process engineering method from that which I recommended. But still, I'll go on to the end.

"Walk in the light, while there is light."

Norm Lieberman
Chemical Engineer
Class of 1964, Cooper Union, New York

Index

A

B

C

D

E

F

G

H

I

J

K

L

M

N

O

P

R

S

T

U

V

W

Other Books by the Author

- *A Working Guide to Process Equipment* (with E.T. Lieberman), 5th Edition (2022) – McGraw Hill Publications
- *Troubleshooting Process Operations,* 4th Edition (2009) – PennWell Corporation
- *Process Engineering for a Small Planet* (2010) – Wiley
- *Process Equipment Malfunctions* (2011) – McGraw Hill Publications
- *Troubleshooting Natural Gas Processing* (2008)– Lieberman Books
- *Process Design for Reliable Operations,* 3rd Edition (2008)– Lieberman Books
- *Process Engineering: Facts, Fiction and Fables* (2017) – Scrivener Publishing
- *Troubleshooting Process Plant Control: A Practical Guide to Avoiding & Correcting Mistakes,* 2nd Edition (2017) – Wiley
- *Troubleshooting Vacuum Systems* (2012) – Wiley
- *Understanding Process Equipment for Operators and Engineers* (2019) – Elsevier
- *Troubleshooting Refinery Processes* (1981) – PennWell Corporation
- *My Race with Death: The Story of One Runner Who's Running Further and Faster With Age* (2013)